# Instructor's Manual and Test Bank

# The Dynamic Earth
### an introduction to physical geology

## Second Edition

## C. Frederick Lohrengel
*Southern Utah University*
*Cedar City, Utah*

## Brian J. Skinner
*Yale University*

## Stephen C. Porter
*University of Washington*

## John Wiley & Sons, Inc.

New York   Chichester   Brisbane   Toronto   Singapore

# PREFACE

This Teacher's Supplement has been written accompany Skinner and Porter's *The Dynamic Earth* 2nd ed. Its purpose is to providing you with a tool that will assist with lecturing introductory level physical geology courses. The included items are: an OUTLINE down to the topical sentence of each paragraph, these are each only one line long; an OVERVIEW containing observations and suggestions concerning that chapter based upon my years of teaching experience; LEARNING OBJECTIVES that cover the topics of each chapter; LECTURE DEMONSTRATIONS that I hope will help liven-up lectures(please share your ideas with me); a list of FILMS AND VIDEO CASSETTES with a list of sources - this list is not exhaustive and is beginning to change rapidly as numerous, relatively inexpensive videos are coming onto the market; a list of REFERENCES - this list is far from exhaustive, but I hope it will give you some ideas.

I hope that you will find these items useful and an aid in your teaching. I hope that you will enjoy this second edition of Skinner and Porter's *The Dynamic Earth*.

C. Frederick Lohrengel II, Ph.D.
Professor of Geology
Southern Utah University
Cedar City, Utah 84720

# CONTENTS

# Part I

# INSTRUCTOR'S MANUAL

# CHAPTER 1

## PLANET EARTH

### OUTLINE

***Vignette of Copernicus and Galileo***
>    Copernicus postulated the planets revolve around the Sun
>    Galileo used the telescope to observe numerous planetary objects

### The Special Planet
>    Atmosphere of gases
>    Hydrosphere of water, ice, water vapor
>    Biosphere of both living and dead organisms
>    Regolith - loose covering of weathered rocks and minerals
>    Life survives by interaction of atmo, hydro, bio and regolith

### The Solar System
>    1 sun, 9 planets, 61 known satellites, assorted small objects
>    Distances between objects in solar system are immense

*Terrestrial Planets*
>    Mercury, Venus, Earth, Mars - small, rocky, dense

*Jovian Planets*
>    Jupiter, Saturn, Uranus, Neptune, Pluto - large, icy, low density

*The Origin of the Solar System*
>    Thinly diffused gases of formed supernova gathered by gravity
>    Gases became dense enough for hydrogen fusion to set in
>    Outer reaches of gas cloud cooled and solid planets condensed
>    Terrestrial planets composed of heavy refractory elements
>    Jovian planets composed of refractory plus volatile elements
>    Planets grew by accretion of many small rocky, icy fragments

*Planetary Accretion: Still Going On!*

      Impact craters -  Meteor Crater, Arizona, 20,000 years

                     Tunguska, Siberia, in 1912

      Impact craters release huge amounts of energy very rapidly

## The Internal Structure of the Earth

      Heated in early history by thermal energy from meteor impacts

      Meteor impact and radioactive heat caused early partial melting

      Earth differentiated while partially melted during early history

*Layers of differing Composition*

      Core - metallic

      Mantle - thick, dense, rocky

      Crust - thin, less dense, rocky

      1. Oceanic crust, average 8 km thick

      2. Continental crust, average 45 km (range 30 to 70 km)

      Seismic wave indicate Earth gets denser with depth

      Seismic waves indicate Earth gets denser in sudden leaps

      Crust and mantle are significantly different in composition

      Core probably iron, based upon meteorite studies

*Layers of Differing Physical Properties*

      Temperature and pressure control physical properties

*The Inner and Outer Core*

      Inner Core solid because of extreme pressure

      Outer core liquid because temperature and pressure balance

      Composition of inner and outer core are both iron

*The Mesosphere*

      Strong, high temperature, high pressure (2883 km to 350 km)

*The Asthenosphere*

      Weak, high temperature, lower pressure (350 km to 100 km)

      Mesosphere and asthenosphere same composition

*The Lithosphere*

      Cooler, stronger, more rigid than asthenosphere(100 km- 0 km)

      Formed from upper mantle and crust

      Layers different composition, similar strength

# Plate Tectonics and the External Structure of the Earth

Lithosphere moved by convection in the asthenosphere
Mountains form as lithospheric plates collide
Ocean basins form where lithospheric plates move apart
Lithosphere is still moving, continents just ride on top
Lithosphere "floats" on top of the moving asthenosphere

## Tectonics

Study of the movement and deformation of the lithosphere
Plate tectonics - the study of lateral movement of lithosphere
Plate tectonic is a basic theme of this text

## Continents and Ocean Basins

Oceans cover 71% of surface area, averages 3.7 km, 11 km max.
Continents cover 29% of surface, average 0.8 km
Continents stand high because of low density - 2.7 gm/cm$^3$
Ocean basins are low because of higher density - 3.2 gm/cm$^3$

## The Shape of Ocean Basins

Continental shelf is flooded continental crust, 25% of total area
Edge of continents is base of the continental slope
Continental crust rocks cover 40% of Earth's surface
Oceanic basin rocks cover 60% of Earth's surface
Continental rise, oceanic crust with thick continental debris pile
Continental margins may or may not coincide with plate margins
All continental margins at sometime in past were plate margins
Nearly as much is known about seafloors as the land surface
Abyssal plains, great flat areas of seafloor adjacent to continents
Oceanic ridges, rocky ridges 84,000 km long
Rift, a narrow, steep valley length of oceanic ridge, volcanism
Trenches, long, narrow, deep basins in seafloor

## Plate Motions

Lithosphere broken into 6 large and many smaller plates
Everything (continent and ocean floor) moves together
Continental drift was forcefully proposed in 1912 by A. Wegener
Original suggestion was only the continents slide over seafloor
Eventually it was discovered that the entire lithosphere moves
Magnetic evidence initially showed continent and seafloor move
Laser beam measurements from satellites shows plates moving

*Plate Margins*

Plate margins are defined by earthquakes and volcanos

Three types of plate margins:

1. Divergent margins or spreading centers
2. Convergent margins with subduction zone or collision zone
3. Transform fault margins

*Spreading Centers*

Plates similar to a broad, irregular conveyor belt

New plate edge is formed at the surface and 2 sides move apart

Magma to form new lithosphere rises from the asthenosphere

Spreading on a continent causes long, linear rifts that drown

Atlantic Ocean started to from 200 m.y. ago and split a continent

Continental slopes on sides of Atlantic match with central ridge

*Subduction Zones*

Near a spreading center the crust is thin

Subduction develops about 1,000 km from the spreading center

Subduction occurs because the crust cools, thickens, is denser

Subducting lithosphere is assimilated into the mantle

Volcanos appear about 150 km from, and parallel to, the trench

*Collision Zones*

Continental crust will never subduct because of low density

Continental crust will collide and form mountains

*Transform Faults*

Zones where two plates slide past each other

Shallow earthquakes common along transform faults

## Interactions Between the Internal and External Layers

Atmosphere, biosphere, hydrosphere break down the crust

*Cyclic Movements*

Materials are added and removed from Earth systems uniformly

Mans activities disrupt the balance of natural cycles

*The Hydrologic Cycle*

Water evaporates, moves, condenses, precipitates, moves to sea

*The Rock Cycle*

*The Three Rock Families*

        Igneous - formed from cooled, consolidated magma

        Sedimentary - formed from lithified precipitates or sediments

        Metamorphic - other rocks altered by great heat and pressure

*Percentage of Rock Types in the Crust*

        Crust is 95% igneous or igneous parented metamorphic rocks

        Sediments are a thin veneer draped over the other 2 types

        The 3 rock types weather, are buried, alter, melt, resurface

*The Cycle*

        The Rock Cycle may take many subpaths during Earth history

        Length of time of a cycle is long - continents average 650 m.y.

        Length of time to cycle the oceanic crust averages 60 m.y.

        Rock and Hydrologic Cycles interact at spreading centers

*Uniformitarianism and Rates of Cycles*

        James Hutton, the same processes have always operated

        19th century geologists thought rates of reactions were uniform

        Evidence is strongly against constancy of rates

        Earth is cooling and internal rates are slowing down

        Cycles are constant, but their rates vary

## A Drill Hole Through the Crust?

        Project Mohole, 1960's, try to drill through crust to mantle

        Project was abandoned because of political, financial, technical

        Deep Sea Drilling Project emerged from the idea, successful

        Continental crust much thicker than oceanic, more problems

        USSR started, and has continued, deep continental drilling

        A hole 45 km deep will probably be drilled eventually

## OVERVIEW

This chapter is an introduction to a significant portion of the topics that will be studied in much greater depth in subsequent chapters. Because of the great variety of topics covered it is a difficult body of material to lecture. Unfortunately quiz grades are typically not good for this chapter and students tend to get discouraged too early in the term and want to drop. As a lecturer your responsibility will be to take these varied topics and use them to excite your students about all of the subjects that will be studied during the next 10 to 15 weeks of the term.

The topics covered begin with an introduction to the Solar System. This will be dealt with more extensively in chapter 18. Differentiation of the planet Earth and surface structures are covered in chapters 15 and 16. The hydrologic cycle is a topic of chapter1. The rock cycle is spread over chapters 1, 3, 4, 5, and 16.

## LEARNING OBJECTIVES

1. Be able to list the 9 planets in the Solar System. Distinguish the terrestrial and jovian planets and why the are separated.

2.Know the method of formation of the planets and when they formed.

3. Know the pattern of differentiation of the planets and the compositional layers of Earth.

4. Know the type of crust and how they differ.

5. Know the layers of Earth based on physical properties and how they differ.

6. Know the number of lithospheric plates and rates of movement.

7. Describe the 3 types of lithospheric plate margins.

8. Describe the interacting cycles between the layers of Earth.

9. Describe the hydrologic cycle and the energy flow.

10. Describe the rock cycle and the interaction within it.

11. Distinguish between the continental and oceanic rock cycles.

## LECTURE DEMONSTRATIONS

See appropriate chapter in this supplement.

## FILMS AND VIDEO CASSETTES

See appropriate chapter in this supplement.

## REFERENCES

See appropriate chapter in this supplement.

# CHAPTER 2

## MINERALS

### OUTLINE

***Vignette on diamonds***

> Objects of great beauty
>
> Formed in mantle at about 150 km, surface in Kimberlite pipes
>
> Very rare, high-grade Kimberlite yields 1 carat per 5 m$^3$ ore

### Minerals and Their Chemistry

> Mineral carries a special connotation
>
> 1. Naturally occurring
>
> 2. Inorganic
>
> 3. Specific composition - which elements and proportions
>
> 4. Specific crystal structure - way the atoms are packed
>
> Rocks are aggregates of minerals, they tell how the world works

*Elements and Atoms*

*Chemical Elements*

> The most fundamental chemical units of matter
>
> Elements identified by 1 capital letter and some also a lowercase
>
> Atom - smallest unit with characteristics of that element

*Atoms and Ions*

> 3 types of components held together by nuclear forces
>
> 1. Proton - large, positive electrical charge
>
> 2. Neutron - large, neutral electrical charge
>
> 3. Electron - very small, negative electrical charge
>
> Atomic number-number of protons in nucleus

*Energy-level Shells*

> Energy - level shells are fixed distances from nucleus
>
> Energy - level shells each have a maximum number of electrons

*Ions*

Energy-level shell with maximum number of electrons is stable

Cation(+), atom that has given up electrons to stabilize shells

Anion(-), atom that has gained electrons to stabilize shells

Superscript number and sign indicate charge and amount

*Compounds*

From when anions and cations combine in specific ratios

Formula - cation followed by anion(s) with subscripts

Molecule is smallest unit of with properties of a compound

*Bonds*

*Ionic Bonds*

Result of electron transfer by the involved ions

*Covalent Bonds*

Result of sharing electrons

Bonding determines the geometric pattern of ions

*Metallic Bonds*

Loosely held covalent bonds on the inner energy-level shells

Outer energy - level shell electrons are free to move if forced

*Van der Waals Bonds*

Weak electrostatic attraction between stable ions or compounds

*Complex Ions*

Anions formed from 2 other ions, act like a single ion however

## Crystal Structure and States of Matter

*States of Matter*

Three states of matter - solid, liquid, gas

Minerals are all solid

State of matter controlled by temperature and pressure

Solid - low temperature and high pressure

Gas - high temperature and low pressure

Liquid - intermediate conditions

*Crystal Structure*

Solids - atoms generally arranged in geometric pattern

Liquids and gases - atoms arranged randomly

Crystal structure - the geometric pattern of atoms in a solid

Amorphous - solids that lack crystal structure

*Ionic Substitution*

> Ionic radius - the distance from the center to outer edge of ions
>
> Anion radii tend to be larger than cation radii
>
> Ionic substitution - random substitution of similar ions in x-alls

## Definition of a Mineral

> For an object to be called a mineral 4 requirements must be met
>
> 1. naturally occurring - eliminates manmade materials
>
> 2. solid - excludes liquids and gases
>
> 3. specific chemical composition - excludes mixtures
>
> 4. specific crystal structure - excludes amorphous materials
>
> Specific chemical composition - constant cation to anion ratios
>
> Mineral group - minerals with extensive ionic substitutions

*Mineraloids*

> Mineraloids are mineral - like but lack all the qualifications

*Polymorphs*

> Compounds that occur in more than 1 crystal arrangement

## The Properties of Minerals

> Determined by composition and crystal arrangement

*Crystal Form and Growth Habit*

*Crystal Form*

> The geometric arrangement of crystal faces
>
> Crystal faces - the planar surfaces that bound a crystal
>
> Crystal - any solid body that grows with planar surfaces
>
> Constancy of interfacial angles - N. Steno, 1669
>
> Internal arrangement of atoms, X-rays, Max von Laue, 1912
>
> Atomic arrangement is constant whether crystal faces or grains

*Growth Habit*

> Every mineral has a characteristic crystal form

*Cleavage*

> Tendency of a mineral to break in preferred directions
>
> Cleavage surfaces are breakages, crystal faces grow
>
> Crystal structure determines cleavages
>
> Many common minerals have distinctive cleavage patterns

*Luster*

 The quality and intensity of light reflected from a mineral

*Color and Streak*

 Striking, but not a reliable property

 Ionic substitution in small quantities will alter colors

*Streak*

 Color of a thin layer of mineral rubbed on unglazed porcelain

*Hardness*

 Relative resistance to being scratched

 Standard is Moh's relative hardness scale(Table 2.2)

*Density and Specific Gravity*

 Density-average mass per unit volume

 Specific gravity-ratio of weight of substance to equal vol. of water

*Mineral Properties and Bond Types*

 Ionic and covalent bonds minerals - hard, brittle

 Metallic and van der Waals bonds mineral - soft, easily deformed

## Common Minerals

 Approximately 3,000 minerals have ben identified

 Only 12 elements occur in crust at more than 0.1% weight

 Many scarce elements commonly occur as ionic substitutions

 Oxygen and silicon make up 70+% of the crust

 Oxygen and silicon combine to produce silicate anion $(SiO_4)^{4+}$

 Silicates most abundant group of minerals in crust

 Oxides second most abundant group of minerals in crust

*The Silicate Minerals*

*The Silicate Tetrahedron*

 4 oxygen ions at 4 corner of a tetrahedron with silicon in center

 Oxygen ions in a silicate tetrahedron may achieve stability by:

 1. ionically bonding with cations

 2. bond covalently with 2 silicon atoms

 Polymerization-linking silicate tetrahedra by oxygen sharing

 Polymerization forms large sheets, chains, circles, 3-D

 Tetrahedra join only at corners, never along edges

*The Olivine Group*

> One group of minerals containing isolated silicate tetrahedra
>
> A fairly abundant group of green minerals
>
> Common in igneous rocks

*The Garnet Group*

> Other group of minerals containing isolated silicate tetrahedra
>
> Common in metamorphic rocks

*The Pyroxene and Amphibole Groups*

> Formed from long chainlike anions
>
> Pyroxenes have general formula - $AB(SiO_3)_2$
>
> > A and B are usually $Mg^{2+}$, $Fe^{2+}$, $Ca^{2+}$, $Mn^{2+}$, $Na^{1+}$, $Al^{3+}$
> >
> > 2 perfect cleavages at 87° and 93°
>
> Amphiboles have general formula - $A_2B_5(Si_4O_{11})_2(OH)_2$
>
> > A is commonly $Ca^{2+}$, $Mg^{2+}$, or $Na^{1+}$
> >
> > B is commonly $Mg^{2+}$, $Fe^{2+}$, or $Al^{3+}$
> >
> > 2 perfect cleavages at 56° and 124°

*The Clays, Micas and Chlorites*

> All formed from polymerized sheets of silicate tetrahedra
>
> All display a pronounced cleavage parallel to the sheets
>
> Clays, most common minerals in regolith
>
> Micas, 2 most common, Muscovite - clear, Biotite - brown
>
> Chlorite, dark green, commonly an alteration product

*Quartz*

> $SiO_2$ only in 3-dimensional network of silica tetrahedra
>
> 6 sided crystals, multicolored due to ionic substitutions
>
> Chalcedony is a fine grained cold water precipitate form
>
> > agate - color banded
> >
> > flint - dark gray or brown
> >
> > chert - light gray or tan
> >
> > jasper - red to orange
> >
> > plasma - dark green

*The Feldspar Group*

> Most common mineral(60%) in Earth's crust
>
> 3-D network of $SiO_2$ plus $Al^{3+}$. $Ca^{2+}$, $K^{1+}$, $Na^{1+}$
>
> Potassium feldspars $K(Si_3Al)O_8$, several polymorphs
>
> Plagioclase$(Na, Ca)(Si, Al)_3O_8$, solid solution series from Ca to Na

*The Carbonate, Phosphate and Sulfate Minerals*
*Carbonates*

Carbonate anion $(CO_3)^{2-}$

3 common minerals-calcite, aragonite, dolomite

Important, common sedimentary(also other 2 groups) minerals

*Phosphates*

Phosphate anion $(PO_4)^{3-}$

Apatite is the common mineral $Ca_5(PO_4)_3(F, OH)$

*Sulfates*

Sulfate anion $(SO_4)^{2-}$

2 common, Anhydrite - $CaSO_4$ and Gypsum - $CaSO_4 \cdot 2H_2O$

*The Ore Minerals*

Minerals that are sought and processed for valuable metals

*Sulfides*

Pyrite-iron, Galena-lead, Sphalerite-zinc, Chalcopyrite-copper

*Oxides*

Magnetite and Hemetite-iron, Rutile-titanium, Cassiterite-tin

## Minerals as Indicators of the Environment of Their Formation

Contain the keys to the conditions under which they formed

Understanding of growth environment from laboratory studies

## Rocks

*Texture*

Overall appearance - size, shape, arrangement of constituents

*Mineral Assemblage*

Varieties and abundances of minerals present

Two levels of study megascopic and microscopic

Microscopic study usually carried out with thin sections

## What Holds Rocks Together?

Igneous and metamorphic grains usually grow together

Sedimentary usually not as well held together

1 - Cemented by mineral precipitated by groundwater

2 - Recrystallization, higher temperature and pressure

## Minerals and Society

       Most abundant minerals in crust have no commercial value, use

       Ore minerals are rare and hard to find

       Can society maintain its utilization level of minerals indefinitely?

       Everything in the Earth's crust is finite and exhaustible

## OVERVIEW

Minerals are the basic units from which rocks and everything else that will be discussed in the remainder of this book. Sadly, minerals are often one of the most difficult to lecture and poorly treated topics. The material sounds like an abbreviated re-hash of a chemistry course. This is a topic basic to the total understanding of geology and you must expend a great effort to excite your students and attract their attention and interest. Any sort of demonstrations, models, stories, etc. you can introduce will help hold their attention and retention of the material.

Be sure to make mineral physical properties and identification an important laboratory exercise, not not invest much lecture time in this. The laboratory should cover 15 to 20 common rock forming and economic minerals.

Much of silicate mineralogy is a study of putting oxygen anions together in different geometric patterns with metallic cations between and ionic substitutions of the metallic cations. Crystal models can help demonstrate these ideas and also relieve some potential boredom.

Minerals also introduce the identification of rocks and the 3 major types of rocks. Be sure to distinguish between rocks and minerals.

## LEARNING OBJECTIVES

1. Be able to list the characteristics that distinguish a mineral from other objects.

2. List and describe the types of bonding, ionic, covalent, metallic, van der Waals.

3. Describe ionic bonding and the basis for its operation.

4. Describe polymorphs and what they are.

5. Know the principal physical properties of minerals and how to use them to identify minerals.

6. Know approximate number of minerals and how many are common to compose most of Earth's crust.

7. Know silicate minerals are most common followed by oxides, carbonates, sulfides, sulfates, and phosphates.

8. Know the basic building block of silicate minerals is the silicate tetrahedron and its structure.

9. Know that the silicate tetrahedrons will polymerize and how.

10. Know the 2 most abundant groups of minerals, feldspars and quartz.

11. Know that rocks can be described in terms of mineral constituents.

## LECTURE DEMONSTRATIONS

1. Crystal structure models are excellent for demonstrating to students the geometrical internal character of crystals. Unfortunately they are usually expensive. Good quality models can be purchased from Klinger Scientific, 83-45 Parsons Boulevard, Jamaica, NY 11432. You can also construct simple models from styrofoam balls and sticks purchased from hobby shops. There are also other sources for models.

2. Large hand samples to show in lecture, or colored slides may be used. A display case with museum quality samples is also exciting for students. In laboratory they will get to handle samples of several minerals.

3. Christman(1980) describes a technique for demonstrating crystal growth using a carousel projector.

4. Schrourd(1977) describes a technique for illustrating the differences in the atomic arrangement of solids, liquids, and gases.

## FILMS AND VIDEO CASSETTES

1. Crystals and Their Structures, 22 min, 1962, Bell Telephone Labs.,(source IU)

2. Minerals (source IV)

3. Minerals and Rocks, 17 min, 1979, (source EB)

## REFERENCES

Asimov, I., 1961, Building Blocks of the Universe. New York: Abelard - Schuman.

Blackburn, W. H., and W. H. Dennen, 1988, Principles of Mineralogy. Dubuque, IA: W. C. Brown.

Bloss, F. D., 1971, <u>Crystallography and Crystal Chemistry: An Introduction.</u> New York: Holt, Rinehart & Winston.

Christman, R. A., 1980, <u>Growing Crystals on the Ceiling.</u> Journal of Geological Education, v. 28, pp. 199-201.

Court, A., and I. Campbell, 1968, <u>Minerals: Nature's Fabulous Jewels.</u> New York: Abrams.

Ernst, W. G., 1969, <u>Earth Materials.</u> Englewood Cliffs, NJ: Prentice-Hall.

Holden, A., and P. Singer, 1960, <u>Crystals and Crystal Growing.</u> New York: Doubleday, Anchor Books.

Hurlbut, C. S., Jr., and C. Klein, 1977, <u>Manual of Mineralogy</u> (after J. D. Dana). 19th ed. New York: Wiley.

Pearl, R. M., 1968, <u>1001 Questions Answered About the Mineral Kingdom.</u> New York: Dodd, Mead.

Schourd, M. L., 1977, <u>Molecular Motion Machine.</u> Journal of Geological Education, v. 25, pp. 22-23.

Zoltai, T., and J. H. Stout, 1984, Mineralogy: Concepts and Principles. Minneapolis, MN: Burgess.

# CHAPTER 3

## THE FIRE WITHIN: VOLCANOS AND MAGMAS

### OUTLINE

**Vignette concerning eruptions of Krakatau and Mount St. Helens**

    August 27, 1883, Krakatau violently erupted and disappeared

    36,000 were killed by the tsunami

    20 km$^3$ of ejecta were put into atmosphere, 5 years to settle out

    Worldwide temperatures dropped 0.5° in 1884

    March 1980, Mount St. Helens in SW Washington activated

    May 18, 1980, the upper 400 m were blown off the volcano

**Magma**

    Molten rock with suspended crystals and dissolved gases

    Volcano - a vent where magma pours out on to the surface

    Lava - magma that pours out on to the surface

    Volcanos are the only places magmas can be studied

    Three conclusions about magmas:

    1. Range of compositions with silica($SiO_2$) predominant

    2. Characterized by high temperatures

    3. Has the properties of a liquid

*Composition*

    Usually expressed in oxide($SiO_2$) form because $O^{2-}$ is so abundant

    Common elements are Si, Al, Fe, Ca, Mg, Na, K, H, and O

    Three distinct types of magmas most common

    1. Basaltic, about 50% $SiO_2$

    2. Andesitic, about 50% $SiO_2$

    3. Rhyolitic, about 60% $SiO_2$

*Gases Dissolved in Magma*

    Present in small amounts(0.2% to 3% by wt.)

    $H_2O$ and $CO_2$ are 98% of gas, also N, Cl, S, Ar

*Temperature*

   Range from 1,000°C to 1,200°C, perhaps to 1,400°C

*Viscosity*

   Flow rates are typically meters per hour or meters per day

   The higher the $SiO_2$ content, the more viscous the magma

   Viscosity is controlled by temperature and composition

*Effect of Temperature on Viscosity*

   The higher the temperature the lower the viscosity

   Pahoehoe, smooth, ropy texture lava is from hot, gassy, fluid lava

   Aa, rubbly, rough textured lava is from cooler, non-gassy, viscous

*Effect of Silica Content on Viscosity*

   The greater the $SiO_4^{4-}$ content, the larger the polymers

   The larger the polymers the more viscous the magma

## Eruption of Magma

   Magma is less dense than surrounding rock and therefore rises

   The higher the pressure, the greater the dissolved gas content

   As magmas rise, pressure decreases, gas releases, bubbles form

*Non-explosive Eruptions*

   Low-viscosity, low gas content magmas tend to be non-explosive

   Spatter cones may form when magma rises and releases gas fast

   Lava tubes from when lava flows from under a hardened crust

   Pahoehoe forms near top and aa near toe of the same flow

   Vesicular lava forms if lava is too viscous for bubbles to escape

   Amygdules are vesicles that have secondary mineral fillings

*Explosive Eruptions*

   Viscous magmas do not allow gas to escape slowly and explode

   Pumice results when gases escape quickly and froth forms

   The higher the viscosity, the greater likelihood if explosions

*Pyroclasts and Tephra*

   Pyroclasts - fragments of rock ejected during volcanic eruption

   Tephra - deposits of pyroclasts

*Eruption Columns and Tephra Falls*

   Explosion of viscous, gas charged magma forms eruption column

   Eruption columns convectively rise as heat is released internally

   Major eruptions may spread ejecta worldwide, decrease temp 1°

*Pyroclactic Flows*
> Hot, highly mobile tephra flows following major eruptions
> Can travel up to 100 km from source, velocities up to 700 km/hr

*Lateral Blasts*
> Form if the side sluffs off, pressure is released, magma explodes

## Volcanos

*Shield Volcanoes*
> Formed by numerous flows of highly fluid lava
> Broad, roughly dome-shaped, slopes of a few degrees
> Flat on top with fluid lava, steeper flanks when more viscous
> Characteristically formed from basaltic lava, little ash or tephra

*Tephra Cones*
> Rhyolitic and andesitic volcanos piling up tephra around vent
> Slopes 25° to 35° depending on size material

*Stratovolcanoes*
> Steep, conical, tephra ≥ viscous lava erupted, interbedded
> 1,000's of m high, 30° slopes at top to 6° to 10° slopes on flanks
> Among the most picturesque volcanos in the world

## Craters, Calderas, and Other Volcanic Features

> Volcanic terrains are unique with fracture, satellite cones, sulfur
> Magma chambers may remain hot for X00,000 of years
> Groundwater seeping down to hot rocks may form geysers

*Craters*
> Open funnel-shaped depression near the top of a volcano

*Calderas*
> Circular, steep opening near top of volcano, several km diameter
> Form when partially emptied magma chamber collapses

*Resurgent Domes*
> Uplift and doming of caldera floor by new magma, new volcanism

*Lava Domes*
> Pile of viscous, low gas lava extruded very late in eruption

*Fissure Eruptions*
> Extrusion of lava along an extended fracture
> Typically basaltic lava, flows into flat lying plateau basalts

*Pillow Basalts*

  Basalt extruded onto seafloor has  surface quenched then cracks

  A quench-crack-quench sequence yields piles of "pillows"

*Volcanic Hazards*

  50 eruptions per year worldwide is typical

  Five types of hazards:

  1. Hot, rapidly moving pyroclastic flows and lateral blasts

  2. Tephra and hot poisonous gases may bury people

  3. Wetted tephra may result in mudflows

  4. Undersea eruptions may cause tsunami

  5. Tephra may devastate agriculture and cause famine

  Since A.D. 1800, 18 eruptions have caused 1,000+ deaths

## Igneous Rocks

  Extrusive igneous rocks form by lava cooling on the surface

  Intrusive igneous rocks form by magma cooling within the crust

  Igneous rock names based on texture and mineral composition

*Texture*

  Extrusive, typically large crystals from slow cooling

  Intrusive, typically small crystals or glass from rapid cooling

  Porphyry, 2 distinct sizes of crystals from 2 phases of cooling

  Glass will form when lava cools too rapidly for mineral formation

  Pegmatites, extremely coarse grained due to high $H_2O$ content

*Mineral Assemblage*

  See Fig. 3.24 for classification chart

  Rocks on left side of Fig. 3.24 are light colored, the right dark

*Varieties of Intrusive Igneous Rock*

  All are coarse grained

*Granite and Granodiorite*

  Feldspar and quartz are chief minerals

  Granite dominated by potassium feldspar

  Granodiorite dominated by plagioclase feldspar

*Diorite*

  Dominantly plagioclase, but with amphibole and/or pyroxene

*Gabbro and Peridotite*

    Gabbro, very dark, > 50% pyroxene and olivine plus plagioclase

    Peridotite, dark, ≥90% olivine

*Varieties of Extrusive Igneous Rock*

    All are fine grained

*Rhyolite and Dacite*

    Both are pale - gray, white, yellow, red, purple

    Rhyolite, potassium feldspar > plagioclase

    Dacite, potassium feldspar < plagioclase

    Obsidian typically rhyolite composition, impurities make it black

*Andesite*

    Dominantly plagioclase, but with amphibole and/or pyroxene

    Typically dirty looking grays, reds, purples, blues, greens

*Basalt*

    Dominate the ocean floors, most common extrusive rock

    Very dark, > 50% pyroxene and olivine plus plagioclase

*Varieties of Pyroclastic Rocks*

    This is a highly variable group of rocks

    Agglomerates are formed from bomb sized tephra

    Tuffs are formed from lapilli and ash

    Mineral assemblage yields remainder of rock name, *i.e.* diorite

    Tephra may become pyroclastic rock in 2 different ways

    1. Cementation by groundwater, most common

    2. Welding(fusing) of hot, glassy ash particles, welded tuffs

## Plutons

    All bodies of intrusive igneous rock, regardless of size or shape

*Minor Plutons*

    Names depend on size and shape of pluton, see Fig. 3.28

*Dikes*

    Tabular, parallel-sided sheet cutting across the host layering

*Sills*

    Tabular, parallel-sided sheet cutting across the host layering

*Laccoliths*

    Mushroom shaped, parallel on base, overlying layers are domed

*Volcanic Pipes and Necks*

      Volcanic neck -circular feeder pipe immediately below a volcano

      Neck - eroded out feeder pipe

*Major Plutons*

*Batholiths*

      Very large, composition varies reflecting differences in magmas

      Length x000 km by width x00 km by thickness 20 to 30 km

      Move upward by stoping and assimilating host rock

      Xenoliths are unassimilated fragments of host rock

*Stocks*

      Irregular shape, <10 km diameter, may be tops of batholiths

## The Origin of Magma

*Distribution of Volcanos*

      Rhyolitic magmas form in the continental crust

      Andesitic magmas from in mantle and rise through anything

      Andesite probably forms from subducting, melting oceanic crust

      Andesite Line parallels subduction zones

      Basaltic magma forms from and in mantle

      Basaltic magma is always erupted from midocean ridges

      Basaltic magma erupts from intraplate volcanos, fed by hot spots

*How Rock Melts*

      Rocks melt at very high temperatures

      James Hutton first to demonstrate rock melting 200 years ago

*Geothermal Gradient*

      The rate at which temperature increases with depth

      Pressure increases with depth

      Geothermal gradient is different beneath continents and oceans

*The Effects of Pressure and Water on Melting*

      Temperature rises to 1000°C at 100 km depth

      Mantle does not melt because pressure at 100 km is 3,500 MPa

      At any given temperature - pressure water(vapor) lowers melting

*Partial Melting*

      Rocks melt over a range because mineral melting points differ

      Magmatic differentiation by partial melting, lowest temp. melt

      Magma with new composition moves, different comp. remains

*The Origin of basaltic Magma*

Basalt minerals are all anhydrous, must come from a dry magma

Basalt composition is produced by 10% - 15% melting of mantle

*The Origin of Andesitic Magma*

May form from complete melting of continental crust

May form from partial melting of wet oceanic crust

The Andesite Line occur 250 km from trenches

*The Origin of Rhyolitic Magma*

Probably of continental origin

1. Rhyolitic volcanos occur on continents or andesitic volc.

2. Rhyolitic magmas are very wet

Partial melting of andesite comp. yields a wet rhyolite comp.

Most rhyolitic magmas cool in crust and produce granite

## Solidification of Magma

Any given composition magma may yield different end products

Minerals freeze at different temperatures, reverse of melting

Each time a suite of minerals is left behind, a new magma moves

Magmatic differentiation by fractional crystallization

*Bowen's Reaction Series*

Fractional crystallization first studied by N.L. Bowen

As crystallization progresses, new minerals progressively form

Magmatic differentiation by fractional crystallization - zoned xalls

Major problems remain concerning fractional crystallization

*Magmatic Mineral Deposits*

Elements not normal in crystals remain after crystallization

Concentrated rare elements are crystallized into pegmatites

Crystal settling concentrates elements early in crystallization

## Volcanos and Geothermal Energy Sources

Geothermal energy is extensively used in Iceland

Most of the Earth's crust is too cool for geothermal energy

Not much energy is readily available from geothermal sources

Much experimentation is currently under way with geothermal

# Overview

The first 2 chapters have gotten the students ready to the understand of the remainder of the course. This chapter begins the fun part. Many of the students will have see volcanos, actively erupting, dormant, extinct. Excitement can be generated by asking about their observations and experiences, give them a few minutes to tell about them unless the class is too large to manage.

With the class excited it is fairly straight forward to proceed through the several topic of the text. Magmas are a good way the reinforce some mineralogy. Types of magmas introduces the nature of eruptions, relate this back to the student experiences. Volcanos and related features grow easily from the eruptive character and again relate back to mineralogy(let them know they lived through that chapter for a reason.)

Types of igneous rocks can be introduced in lecture(classification scheme) and then expanded in laboratory, samples or videos help immensely. Plutons require some sort of video assistance.

Origin of magmas is an imagination exercise that can be related to cooking experiences the students have had. Cooking helps with Solidification of magmas. Think into your own experiences for examples.

## LEARNING OBJECTIVES

1. List the principal controls on the physical properties of magma.

2. Know the relationship between chemical composition and fluidity of magmas.

3. Know the three types of magmas.

4. Know how partial pelting produces different types of magmas.

5. Describe how shape of volcanos is controlled by composition of magma.

6. Know magma forms by complete or partial melting of rocks.

7. Know igneous rocks form by crystallization of magmas.

8. Know igneous rocks form either on the surface or within the crust of the Earth.

9. Know the texture of igneous rocks is controlled by where the magma cools.

10. Know where various compositions of igneous rocks are most likely to form.

11. Know intrusive igneous rocks bodies are called plutons.

12. Know the names of the several types of plutons and be able to describe them.

13. Describe the distribution of volcanic activity and how the different compositions are distributed.

14. Know that hot spots occur in the interiors of plates and the types of structures they produce.

15. Describe the order of crystallization of minerals in a magma.

16. Describe how molten rock is separated from already formed crystals.

17. Describe magmatic mineral deposits.

## LECTURE DEMONSTRATIONS

1. Large hand specimens may be shown during lecture, or in very large classes slides may be used. In laboratory the students should handle small samples and classify them.

2. Photomicrographs and photographs of igneous textures and features may be purchased from scientific supply houses or you may made your own.

4. Three-dimensional models of intrusions and volcanos can be purchased or you can make our own from wood, styrofoam, plaster of paris, etc. Make the cut away models.

## FILMS AND VIDEO CASSETTES

1. <u>Rocks that Originate Underground</u>, 22 min., 1966, Encyclopedia Britannica (source: EBE).

2. <u>Eruption: St. Helens Explodes</u>, 27 min., 1980, Educational Media Center, (source: EM).

3. <u>The Rock Cycle</u>, 22 min., 1982, Encyclopedia Britannica (source: EBE).

4. <u>Lavas of Etna</u>, 24 min., 1982, (source: UM).

5. <u>Volcanos: Exploring the Restless Earth,</u> 18 min., 1973, Encyclopedia Britannica (source: EBE).

6. <u>Eruption of Kilauea, 1959-60,</u> 28 min., 1960, (source: MTP)

7. <u>Fire Under the Sea,</u> 20 min., 1974, (source: MP)

8. <u>Volcano Surtsey,</u>25 min., 1966, (source: GIS)

9. <u>An Active Volcano in Eruption,</u> 26 min., 1970, (source: GIS)

10. <u>Fire on Heimaey,</u> 33 min., 1975, (source: VS)

11. <u>Heartbeat of a Volcano,</u>21 min., 1970, (source: EB)

12. <u>Fire Mountain,</u> 9 min., 1970, (source: EB)

13. <u>Volcano: Birth of a Mountain,</u> 24 min., 1977, (source: EB)

14. <u>The Violent Earth,</u> 52 min., 1973, (source: NGS)

15. <u>Lavas of Etna,</u> 24 min., 1982, (source: OU)

16. <u>Anatomy of a Volcano,</u> 52 min., 1980, (source: TLV)

17. <u>Volcanos,</u> 28 min., 1985, (source: PBS)

18. <u>Pèlè's March to the Sea,</u> (source: TV)

## REFERENCES

Best, M. G., 1982, <u>Igneous and Metamorphic Petrology.</u> San Francisco: Freeman.

Blong, R. J., 1984, <u>Volcanic Hazards,</u> Orlando, FL: Academic Press.

Bolt, B. A., W. C. Horn, G. A. MacDonald, and R. F. Scott, 1977, <u>Geological Hazards,</u> New York: Springer-Verlag.

Bryson, R. A., and B. M. Goodman, 1980, <u>Volcanic Activity and Climatic Changes,</u> Science, v. 207, pp.1041-1044

Bullard, F. M., 1976, <u>Volcanos of the Earth, 2nd ed.,</u> Austin, TX: University of Texas Press

Cox, K.G., J. D. Bell, and R. J. Pankhurst, 1979, <u>The Interpretation of Igneous Rocks,</u> Boston: Allen and Unwin.

Crandell, D. R., and D. R. Mullineaux, 1978, <u>Potential Hazards from Future Eruptions of Mount St. Helens Volcano,</u> U. S. Geological Survey Bulletin 1383-C

Decker, R., and B. Decker, 1981, <u>The Eruption of Mount St. Helens,</u> Scientific American, v. 244, no. 3, pp. 68-80.

Ernst, W. G., 1969, Earth Materials. Englewood Cliffs, NJ: Prentice-Hall.

Fisher, R. V., and H. U. Schmineke, 1984, Pyroclastic Rocks. New York: Springer-Verlag.

Foxworthy, B. L., and M. Hill, Volcanic Eruptions of 1980 at Mount St. Helens, the first 100 Days. Washington, D.C.:U. S. Geological Survey Professional Paper 1249.

Francis, P., 1976, Volcanos. New York: Penguin Books.

Gilluly, J., 1971, Plate Tectonics and Magma Evolution. Geological Society of America Bulletin, v. 82, pp. 2383-2396.

Greeley, R., 1974, Hawaiian Planetary Conference. Washington, D.C.: NASA.

Green, J., and N. M. Short, 1971, Volcanic Landforms and Surface Features. New York: Springer-Verlag.

Hyndman, D. W., 1972, Petrology of Igneous and Metamorphic Rocks. New York: McGraw-Hill.

Lipman, P. W., and D. R. Mullineaux (eds.), 1981, The 1980 Eruptions of Mount St. Helens. Washington, D.C.: U. S. Geological Survey Professional Paper 1250.

McBirney, A. R., 1984, Igneous Petrology. San Francisco: Freeman, Cooper and Co.

Sigurdsson, H., 1985, The Eruption of Vesuvius in A.D. 79. National Geographic Research, v. 1, no. 3, pp. 332-387.

Simkin, T., and R. S. Fiske, 1983, Krakatau. Washington, D. C.: Smithsonian Institution.

Williams, H., and A. R. McBirney, 1979, Volcanology. San Francisco: Freeman, Cooper, and Co.

Williams, H., F. J. Turner, and C. M. Gilbert, 1982, Petrology (2nd Ed.). San Francisco: W. H. Freeman.

# CHAPTER 4

## SEDIMENTS AND SEDIMENTARY ROCKS

### OUTLINE

#### *Vignette concerning depositional sites*

    Sediment eroded from the lands finally ends up in the oceans

    Deep sea drilling has yielded very complete sediment sequences

    Deep sea exploration was fundamental to plate tectonics

    Numerous spin offs from the deep sea drilling project

#### **Sediments and Sedimentation**

    Sediment is nearly everywhere

*Stratification and Bedding*

    Arrangement of particles in layers produces stratification

    Layering occurs in volcanic and metamorphic rocks

    The layered arrangement of strata is refereed to as bedding

    The top or bottom of a bed is a bedding plane

*Clastic Sediment*

    An individual fragment of rock or mineral is a clast

    Mechanical breakdown of older rocks produces detritus

    Size range of detritus is large boulders to submicroscopic clays

    Table 4.1 shows range of clast sizes

*Production of Clastic Sediment*

    Clasts are produced by weathering of older rocks

    Detritus may also be produced from skeletal parts or shells

*Transportation and Deposition of Clastic Sediment*

    May move by gravity, glaciers, wind, running water

    Deposition occurs when transporting agent looses energy

    Greater the energy of the transporting agent, larger clasts move

*Mineral Composition of Clastic Sediment*

Coarse clastic are dominated by resistant rocks and minerals

The longer transportation goes on the more breakdown occurs

Eventually quartz and clay remain after enough cycles

*Some Conspicuous Features of Clastic Sediments*

Size, packing, sorting are important

*Sorting*

A measure of the range of particle size in sediments

Grain sizes changes are result of fluctuations of energy of agent

Specific gravity also effects sorting

Sorting is typically by size because specific gravities are similar

Sorting is also by durability, nondurable are destroyed

*Particle Shape*

Clasts are initially angular

Transportation causes rounding and sphericity of clasts

*Rhythmic Layering*

Any distinctive alteration of parallel bedding

Varves - summer/winter alterations

*Cross Bedding*

Inclined layering within the major stratum

*Graded Bedding*

Various origins, coarsest clasts at the base getting finer upward

*Nonsorted Sediment*

Chaotic mixture of sizes

Typically mass wasted or glacial

*Chemical Sediment*

Material that was transported in ionic solution

1. Biochemical activities of plants and animals

2. Evaporation and precipitation from lakes or ocean

*Biogenic sediment*

Sediment composed mainly of fossil remains whole or fragments

*Calcareous and Siliceous Biogenic Sediment*

Most calcium carbonate precipitation involves an organism

Deep-sea ooze, mix of mud and tests of dead surface organisms

Calcareous ooze - foraminifers and coccolithophores

Siliceous ooze - radiolarians and diatoms

*Growth and Burial of Plant Matter*

Photosynthesis, plant conversion of solar into chemical energy

About 1% of buried organic does not decompose, stays trapped

*Fossil Fuels*

Coal, oil, natural gas

Vary with original organics and sediment type plus later events

*Peat*

Unconsolidated deposit of plant remains, carbon content >60%

*Petroleum*

Altered marine microphytoplankton & bacteria trapped in mud

Evidence that oil originates from decomposition of organics

1. possesses optical properties from organic hydrocarbons

2. contains nitrogen and other compounds common to living

Conversion to petroleum is long and complicated process

## Lithification and Diagenesis

Lithification - conversion of sediment to sedimentary rock

Diagenesis - any type of change to sediment after deposition

1. simplest form is compaction

2. cementation with minerals deposited by groundwater

3. recrystallization is forming polymorphs or size increase

4. oxidation or reduction of organics

Plant material to peat to lignite to coal is a diagenetic series

## Common Sedimentary Rocks

Clastic, chemical, biogenic

*Clastic Sedimentary Rocks*

Sedimentary rocks, transported fragments cemented together

Clastic sedimentary rocks are classified on basis of size clast

Conglomerate - gravel, cobbles, boulders >2 mm

| | | |
|---|---|---|
| Sandstone ----- sand | | 2 - 1/16 mm |
| Siltstone ------- silt | | 1/16 - 1/256 mm |
| Shale ----------- mud | | <1/256 mm |

Conglomerate large, rounded rock fragments w/ smaller clasts

Breccia large, angular rock fragments w/ smaller clasts

Sandstone, dominated by sand sizes clasts
1. Quartz sandstone - dominated by quartz grains
2. Arkose - feldspars a major component
3. Lithic Ss - rock fragments a major component

Siltstone, silt sized clasts, predominantly quartz and feldspar

Shale, clay sized, clay minerals, quartz, feldspar, calcite

*Chemical Sedimentary Rocks*

Lithification of organic or inorganic chemical precipitates

*Common Rock Types*

Rock salt - halite(NaCl)

Gypsum - gypsum($CaSO_4 \cdot 2H_2O$)

Chert - amorphous fine-grained silica, some with organic tests

*Mineral Deposits in Chemical Sedimentary Rocks*

Contain most of world's minerals and energy resources

*Iron Deposits*

Most of Earth's iron ore is in ancient chemical precipitates

*Phosphorus Deposits*

Apatite [$Ca_5(PO_4)_3(OH, F)$] precipitates from seawater

*Evaporite Deposits*

Formed when a lake or arm of the ocean evaporates away

Numerous minerals form, many are hydrated

*Biogenic Sedimentary Rocks*

Lithified biogenic or high organic content sediments

*Limestone and Dolostone*

Dominated by minerals calcite and dolomite

Numerous textures reflecting varied origins and sources

*Diatomite*

Lithified diatom and radiolarian tests

*Coal*

Black, combustible sedimentary rock, >50% decomposed plants

Plant material » peat » lignite » bituminous coal, increased burial

Mostly accumulates in tropical to subtropical swamps/marshes

*Oil Shale*

Shale with unusually high content of organics - kerogens

*Environmental Clues in Sedimentary Rocks*

        Sum total of all features are required to make a conclusion

*Features on Bedding Planes*

        Sole marks - directional and current marks on bedding planes

        Ripple marks - small cross-bedded structures

        Mudcracks - features formed when mud dries out and shrinks

        Trace fossils - tracks, trails, burrows, borings, all are organic

        Raindrop impressions - from a brief, intense rain shower

*Fossils*

        Provide significant clues to ancient environments

*Color*

        Dominated by colors of minerals, rock fragments, organic matter

        Dark colors - iron sulfides and organics

        Reds, browns, and yellows - iron oxides

        Whites and grays - common rock forming minerals

        Weathering often changes the rock color as new minerals form

## Sedimentary Facies and Depositional Environments

        Layers of sediment are similar to pages of a book

*Sedimentary Facies*

        Strata change appearance reflecting depositional environments

        On any time plane a large number of environments exist

        Facies are distinguishable depositional environments

        Facies studies can give a picture of the original environments

*Common Sediments of Nonmarine Environments*

        En route to the sea, numerous temporary depositional sites exist

*Stream Sediments*

        Principal agency for sediment transportation

        Numerous localized environments of deposition

*Lake Sediments*

        Generally well sorted mud, sand, and gravel, fresh-water fossils

*Glacial Sediments*

        Mixtures of random sized clasts

*Eolian Sediments*

        Well sorted, size and mineralogy, generally extensive deposits

*Sediments of the Continental Shelves*

Most sedimentary rocks accumulated on the shelves

*Estuarine Sediments*

Semi-enclosed, brackish water, coarse to fine clasts seaward

*Deltaic Sediments*

Mass of sediment where a stream enters the ocean or a lake

*Beach Sediments*

Linear deposits, well sorted and rounded, cross stratified

*Offshore Sediments*

Generally very fine sediment in thin, uniform blankets

Recent sealevel changes have left coarse deposits far from shore

90% of the continental detritus is deposited on the shelves

*Carbonate Shelves*

Warm climate and sea-surface temp. and little land detritus

*Marine Evaporite Basins*

Restricted circulation ocean basins in hot, dry climate

Wide spread in the geologic record, underlie 30% of No. Amer.

*Sediments of the Continental Slope and Rise*

The "steep" slope to the oceanic depths

*Turbidity Currents and Turbidites*

Gravity-driven currents of dilute water-sediment mixtures

Turbidity currents are extremely effective transportation agents

Turbidites are turbidity current deposits, graded bedding

Turbidites are deposited infrequently at any single spot

*Deep-Sea Fans*

Land derived sediment deposited at the base of submarine cyns.

Major exceptions that most sediment is trapped on shelves

*Sediment Drifts*

Huge deep ocean currents deposits, base of continental margin

*Sediments of the Deep Sea*

These are mixtures of surface biogenic and continental material

*Deep-Sea Oozes*

Calcareous ooze in low and mid latitudes <4km deep

Siliceous ooze in higher latitudes and >4km deep

*Land-derived Sediment*

Eroded and transported by a variety agents and sources

## Sedimentation and Tectonics

Energy to drive sedimentation - the sun, radioactivity, gravity

Sedimentation greatest near active tectonism decreases away

Great accumulations common on trailing continental edges

Basins near continental collision zones gain great thicknesses

Subducting regions typically have high percentage of volcanics

## Geology and the Politics of Middle East Oil

Petroleum is not uniformly distributed over globe

Petroleum bearing rocks are concentrated in relatively few spots

Several small Persian Gulf countries posses most of world's oil

Millennia of intertribal and religious conflict now include oil

## OVERVIEW

Sedimentary rocks are the thin veneer that covers most of the continents. Sediments cover a significant portion of the seafloors. It is from the study of sedimentary rocks that much of the history of the Earth has been unraveled. These are the sedimentary rocks that a large portion of the geologists spend their careers. Sedimentary rocks are the ones most students will be most familiar with.

This chapter deals with the processes of formation and deposition as well as the lithified rocks. The processes of deposition are also areas that many of the students will be familiar and will be able to visualize.

Sedimentary rocks are of interest to students because they contain all of the fossil fuel supplies.

This chapter will lend itself well to the use of visual aids, slides, specimens, etc.

## LEARNING OBJECTIVES

1. Know that sediment is transported by streams, glaciers, wind, slope processes, and ocean currents.

2. Know that after deposition sediment experiences compaction and cementation as it is transformed onto sedimentary rock.

3. Know that stratification results from the arrangement of sedimentary particles in layers.

4. Know each sedimentary bed in a succession of strata is distinguished by its distinctive thickness or character.

5. Know that clastic sediment consists of fragmental rock debris resulting from weathering plus broken remains of organisms.

6. Know that chemical sediment forms where substances transported in solution are precipitated.

7. Know that particles of sediment become rounded and sorted during transportation by water and wind, but not by glaciers.

8. Know that clasts have different arrangements in strata, parallel and cross strata, graded bedding, nonsorted layers.

9. Know that oil and gas probably originated as organic matter deposited in seafloor sediments and were then decomposed chemically and were then trapped as fluids at a later time in a geological trap.

10. Know that oil shales are a major potential source of fossil fuel, but that they must be heated to release the oil and gas.

11. Know coal originated as plant matter in ancient swamps, and that it is both abundant and widely distributed.

12. Know that sediments are transformed into sedimentary rocks during diagenesis by the processes of compaction, cementation, and recrystallization.

13. Know that both clastic sedimentary rocks and clastic sediments are classified on the basis of size clasts. Conglomerate, sandstone, siltstone, shale are rock equivalents to gravel, sand, silt, clay.

14. Know that limestone is a widespread biogenic rock that forms primarily in warn marine environments and that it stores carbon dioxide in the earth's crust.

15. Know that an aerially extensive body of strata may posses several facies, each determined by a different depositional environment. Facies boundaries may be abrupt or gradational.

16. Know that most sedimentary strata are composed of continental detritus that was transported and deposited on the submerged continental shelf.

17. Know that some strata represents detritus trapped in nonmarine continental basins.

18. Know that only a small amount of continental detritus is transported to deep marine.

19. Know that depositional environments of nonmarine and shallow marine sediments can be inferred from such properties as texture, degree of sorting and rounding, character of stratification, and types of contained fossils.

20. Know that coarse land-derived sediment reaching the continental margins is deposited close to shore where it is reworked by longshore currents.

21. Know that fine sediment is deposited on the continental shelves and slopes and in the deep sea.

22. Know that extensive areas of the shelves are covered by relict sediments deposited at times of lower sea level.

23. Know that carbonate shelves are found in low latitudes where warm waters promote growth of carbonate-secreting organisms.

24. Know that evaporite deposits accumulate in restricted marine basins where evaporation is high and continuous inflow provides a supply of saline water.

25. Know that turbidity currents have built large deep-sea fans at the base of the continental slope be the deposition of turbidites.

26. Know that thew chief kinds of sediment on the deep seafloor are brownish or reddish clay, calcareous ooze, and siliceous ooze.

27. Know that the distribution of deep-sea deposits is related to surface water temperature, water depth, and surface productivity.

28. Know that sediment is constantly being recycled, moving from continent to ocean and back to the continent.

## LECTURE DEMONSTRATIONS

1. If there is a stream table available many sedimentary feature can be built.

2. Graded bedding can be produced nicely if a 1000 ml graduated cylinder. A little experimentation with the amount of material will yield desirable results.

3. The is a chapter where lots of slides of rock types, environments, and structures will greatly aid the students.

4. Large hand specimens plus smaller ones in laboratory will produce a decided return in enhanced learning.

## FILMS AND VIDEO CASETTES

1. Rocks That Form on the Earth's Surface, 16 min., 1964, (source: EB, BU, IU, USC).

2. Carbonate Sedimentation, Florida Shelf Margin, 45 min., 1975, (source: SOC).

3. The Rock Cycle, 22 min., 1982, (source: EBE).

4. The Formation of Sedimentary Rock, (source: IV).

5. Sedimentary Rocks, 18 min., (source: MLA).

6. Great Barrier Reef, 26 min., 1968, (source: AIS)

7. Fossils: Exploring the Past, 16 min., 1978, (source: EB)

## REFERENCES

Bathhurst, R. G. C., 1975, Carbonate Sediments and Their Diagenesis, 2nd ed. New York: Elsevier.

Davis, R. A., 1983, Depositional Systems. Englewood Cliffs, NJ: Prentice-Hall.

Dunbar, C. O., and J. Rogers, 1857, Principles of Stratigraphy, New York: Wiley.

Freedman, G., and J. Sanders, 1978, Principles of Sedimentology. New York: Wiley.

Galloway, W.E., and D. K. Hobday, 1983, Terrigenous Clastic Depositional Systems. New York: Springer-Verlag.

Krumbein, W. C., and L. L. Sloss, 1963, Stratigraphy and Sedimentation, San Francisco: W. H. Freeman

Pettijohn, F. J., 1975, Sedimentary Rocks. 3rd ed. New York: Harper and Row.

Pettijohn, F. J., P. E. Potter, and R. Sevier, 1972, Sand and Sandstone. New York: Springer-Verlag.

Reineck, H. E., and I. B. Singh, 1975, Depositional Sedimentary Environments, New York: Springer-Verlag.

Rigby, J. K., and W. K. Hamblin, 1972, <u>Recognition of Ancient Sedimentary Environments.</u> Tulsa, OK: Society of Economic Paleontologists and Mineralogists, Special Publication 16.

# CHAPTER 5

## NEW ROCKS FROM OLD:
## METAMORPHISM AND METAMORPHIC ROCKS

### OUTLINE

***Vignette on the character of metamorphic rocks***

> Metamorphism, all mineral assemblage and rock texture change
> Metamorphism, changes caused by temperature and pressure
> Record of each metamorphic event is preserved in the rocks
> Much tectonic history is preserved in metamorphic rocks

### The Limits of Metamorphism

> Temperatures > 200°C and pressures > 300 MPa and only solids
> Wet partial melting causes igneous processes to set in
> Migmatites - mostly metamorphic, but with some melting
> Granitic batholiths with voluminous metamorphics is common
> Ig - Met rock association common to subduction - collision zones
> Low-grade metamorphism - 200° to 320°C and low pressure
> High-grade metamorphism - > 550°C and high pressure

### Controlling Factors in Metamorphism

> Initial chemistry, conditions of metamorphism controls product
> *Chemical Reactivity Induced by Fluids*
> Intergranular fluids play a vital role in metamorphism
> Composition of intergranular fluids change with temp. - press.
> Speed of reactions increase with increasing intergranular fluids
> Pressure decreases porosity and amt. of hydrous minerals
> Prograde metamorphic effects - rising temp. and press.
> Retrograde metamorphic effects - decreasing temp. and press.
> Retrograde much slower than prograde

*Pressure and Temperature*

    Changes of temperature and pressure occur simultaneously
    Laboratory work suggests temp. - press. conditions at formation

*Differential Stress*

    Metamorphic textures record differential stress of formation
    Differential stress produces directional oriented minerals
    Homogeneous stress produces randomly oriented minerals

*Time*

    Two compounds react only if product is at a lower energy state
    Metamorphic reactions typically take millions of years
    Coarse grains - high temp. and press. and long times
    Fine grains - lower temp. and press. and shorter times

## Metamorphic Responses to Changes in Temperature and Pressure

*Textural Responses*

    Most metamorphic rocks develop conspicuous directional text.
    Foliation - planar texture caused by parallel oriented grains

*Slaty Cleavage*

    Property of low-grade met. rocks breaking into platelike frags.
    Caused by mineral growth perpendicular to stress direction

*Schistosity*

    Cleaves easily, visible coarse grains, arranged parallel, sheet-like

*Assemblage Responses*

    As temp. - press. increase, mineral assemblages change, Fig5.7
    Metamorphic minerals - chlorite, serpentine, epidote, talc, etc.

## Kinds of Metamorphic Rocks

    Names are based upon both texture and mineral assemblage

*Metamorphism of Shale and Mudstone*

*Slate*

    Lowest grade metamorphism of shale or mudstone is slate
    Slaty cleavage develops as muscovite and chlorite crystallize

*Phyllite*

    Grains just visible to unaided eye, pronounced foliation

*Schist and Gneiss*

      Schist - coarse-grained with pronounced schistosity

      Gneiss - coarse-grained, pronounced foliation, segregated min.

      Schist and gneiss - textural names with mineral name adjectives

*Metamorphism of Basalt*

*Greenschist*

      Chlorite gives green color to low-grade rock with hydrous mins.

*Amphibolite and Granulite*

      Amphibole replaces chlorite at intermediate grade, less foliated

      Pyroxene replaces amphibole at high grade, very poor foliation

*Metamorphism of Limestone and Sandstone*

      Limestone—>marble and sandstone—>quartzite, lack foliation

*Marble*

      Structureless mass of coarse, crystalline, interlocking calcite

*Quartzite*

      Formed by infilling pores with silica and mass recrystallization

## Kinds of Metamorphism

      Mechanical deformation - grinding, crushing, dev. of foliation

      Chem. recrystallization - comp., new mins., loss of $H_2O$, $CO_2$

*Cataclastic Metamorphism*

      In brittle rocks with intense, differential stress

      Individual grains are shattered, pulverized, elongated, foliated

*Contact Metamorphism*

      Adjacent to hot magmas chemical recrystallization, little mech.

      Develop an aureola, width dependent on intrusive size and $H_2O$

      Aureola is zones by temperature and water content

*Burial Metamorphism*

      Produces met. rocks that look sed., but new mineral assemblage

      Zeolites most common, feldspar chemistry with water

      Common in deep tectonic basins and marginal trenches

      Grades into regional metamorphism

*Regional Metamorphism*

      Very wide spread, distinctly foliated, differential stress

      Form in subduction and collision zones result of plate tectonics

      Different products are produced depending on temp. and press.

*Metamorphic Zones*
> First studied in Scottish Highlands
> Isograd, line connecting points of equal metamorphic intensity
> Metamorphic zones are regions between isograds

## Metamorphic Facies
> Only minor chemical changes occur during metamorphism
> Temperature and stress determine the final metamorphic mins.
> Metamorphic facies - equilibrium minerals, specific conditions
> When comparing metamorphic facies be sure chemistry is same

## Metasomatism
> Most metamorphic activity has water:mineral ratios of 1:(>10)
> Metasomatism involves major ionic alteration involving water
> Metasomatism is commonly associated with contact meta.

*Hydrothermal Solutions and Mineral Deposits*
> Metasomatism involved hydrothermal solutions - water 250+°C
> Probably more mineral deposits are hydrothermal than any other

*Composition of the Solutions*
> Most hydrothermal solutions are 3.5+% dissolved salts
> Minerals are much more soluble in brines than pure water

*Origins of the Solutions*
> Cooling and crystallization of wet partial met magmas
> Often associated with volcanism
> Submarine hydrothermal springs occur along spreading centers

## Metamorphism and Metasomatism
> Regional metamorphism occurs at the subducting plate boundary
> Burial metamorphism occurs at base of thick piles of sediment
> Blueschist and eclogite facies, rapidly subducting plates
> Greenschist and amphibolite facies, continental collision zone
> Metasomatism/hydrothermal, regional meta. + magmatic activity

# Pressure-Temperature-Time Paths for Metamorphic Rocks

Metamorphic grade produces a distinctive mineral assemblage

Meta. mins. often have relict minerals from earlier assemblages

Subduction rates can be determined from P-T-t paths

## Overview

Metamorphic rocks are possibly the most interesting of the three groups, but at the same time the most difficult to understand. They are abundant, particularly in the core of mountain ranges. They constitute a significant portion of the craton and underlie much of the thin veneer of sedimentary rocks on the continents. The difficult part of metamorphic rocks is that they form in an environment that is completely unfamiliar to most people.

This chapter includes discussions on the agents of metamorphism(pressure, heat, solutions), types of metamorphism(local and regional), types of rocks produced. Regional metamorphism is related to mountain building, local metamorphism is related to magmatic activity. many economic deposits are related to metamorphic activity, an important point to emphasis.

It is difficult for most students to grasp the not melting occurs and that all of the changes occur in a solid state in the presence of small amounts of water. Another point is that each mineral has a range of temperature and pressure over which it is stable. Some analogies to gradational or regional metamorphism is a cake baking in an oven that is not uniformly heated and one end is burned and the other end is still raw.

## LEARNING OBJECTIVES

1. Know that metamorphism involves changes in mineral assemblages and rock textures and occurs in the solid state as a result of changes in temperature and pressure.

2. Know that mechanical deformation and chemical recrystallization are the processes that affect rock during metamorphism.

3. Know the presence of intergranular fluid greatly increases the rate of metamorphic reactions.

4. Know that foliation, as expressed by directional textures such as slaty cleavage and schistosity, arises from parallel growth of minerals formed during metamorphism.

5. Know that cataclastic metamorphism involves mechanical deformation but little or no chemical recrystallization.

6. Know that heat given off by bodies of intrusive igneous rock causes contact metamorphism and creates contact metamorphic aureoles.

7. Know that contact metamorphism involves chemical recrystallization but very little mechanical deformation.

8. Know that regional metamorphism is the result of regional metamorphism and involves both mechanical deformation and chemical recrystallization.

9. Know that regionally metamorphosed rocks are produced along subduction and collision edges of plates.

10. Know that if rocks of the same chemical composition are subjected to the same metamorphic conditions, the same mineral suite is produced.

11. Know that each mineral assemblage defines a metamorphic facies, and that each facies forms under a specific range of stress and temperature.

12. Know that metasomatism involves the changes in rock composition that occur when material in solution is added to the rock, or taken away, as the result of fluids flowing through a rock.

13. Know that hydrothermal solutions are naturally formed hot brines that are capable of dissolving, transporting, and precipitating minerals.

14. Know that metamorphism can be explained by plate tectonics.

15. Know that burial metamorphism occurs within the thick piles of sediment at the foot of continental slopes.

16. Know that regional metamorphism occurs in regions of subduction and continental collision.

17. Know that metasomatism due to hydrothermal solutions is linked to plate tectonics because the solutions tend to form in, or be associated with, stratovolcanoes.

18. Know that stratovolcanoes are formed above subduction zones.

## LECTURE DEMONSTRATIONS

1. Large specimens and/or slides of metamorphic rocks is indispensable for demonstrating the character of metamorphic rocks.

2. Metamorphic grade can be demonstrated by either using the specimens in hand samples or through slides in increasing grade.

3. Alignment of platy minerals can be demonstrated by piling up a bunch of wads of different colored paper(in a clear contained) then pressing them flat. The wads will flatten into platy discs with slaty cleavage.

## FILMS AND VIDEO CASSETTES

1. <u>Rocks that Originate Underground,</u> 23 min., 1966, (source: EB).

2. <u>Metamorphic Rocks,</u> (source: IV).

## REFERENCES

Atherton, M. P., and C. D. Gribbe, ed., 1983, <u>Migmatites, Melting, and Metamorphism,</u> Cheshire, U. K.: Shiva Publications.

Barth, T. F. W., 1962, <u>Theoretical Petrology 2nd ed.,</u>New York: Wiley.

Best, M. G., 1982, <u>Igneous and Metamorphic Petrology,</u> San Francisco: Freeman.

Ernst, W. G., ed., 1975, <u>Metamorphism and Plate Tectonic Regimes: Benchmark Papers in Geology,</u> Stroudsburg, PA: Dowden, Hutchinson and Ross.

Ernst, W. G., ed., 1975, <u>Subjection Zone Metamorphism: Benchmark Papers in Geology,</u> Stroudsburg, PA: Dowden, Hutchinson and Ross.

Hyndman, D. W., 1972, <u>Petrology of Igneous Rocks,</u> New York: McGraw-Hill.

Mason, R., 1978, <u>Petrology of the Metamorphic Rocks,</u> Boston: Allen and Unwin.

Menhart, K. R., 1968, <u>Migmatites and the Origin of Granitic Rocks,</u> Amsterdam: Elsevier.

Turner, F. J., 1968, <u>Metamorphic Petrology,</u> New York: McGraw-Hill.

Vernon, R. H., 1976, <u>Metamorphic Processes,</u> Boston: Allen and Unwin.

# CHAPTER 6

## GEOLOGIC TIME

### OUTLINE

**Vignette on relative and absolute time**

Relative time is the sequence of past events

Absolute time is the time in numbers of years

James Hutton was the first to appreciate significance of rel. time

Charles Lyell demonstrated determination of relative time

Radioactivity discovered in 1896, geologic time immense

**Stratigraphy**

Earth history reconstructed from sequence and relative ages Sed

Study of strata is stratigraphy

*Original Horizontality*

Each new layer of sed. rock is laid down horizontally over older

*Stratigraphic Superposition and the Relative Ages of Strata*

Principle of Stratigraphic Superposition, strata get younger up

In disturbed sed. sequences, the top must be determined

*Breaks in the Stratigraphic Record*

Lyell- dated Earth from thickness of sed. rock vs rate of accumul.

Unconformity - a substantial break in a stratigraphic sequence

*Kinds of Unconformities*

Angular unconformity, angular discordance above and below

Disconformity, erosion surface between parallel sequences

Nonconformity, strata on eroded igneous and metamorphics

See Figs. 6.5, 1, and 2

Unconformities relate tectonics, erosion, and sedimentation

Accumulation in one place compensates for erosion in another

## Stratigraphic Classification

Strata depict the physical - biological character of place in time
Rock stratigraphic unit, distinctive from rock above and below
Formation -  distinctive rock unit that can be mapped
Time-stratigraphic unit, all rock deposited during a given time
Formation boundaries transgress through time and space
System, primary time-stratigraphic unit, worldwide in extent
Nonmaterial units are the time during which rocks accumulate
Period is the primary unit of geologic time
Correlation, piecing together a complete strat. sequence

## Correlation of Rock Units

Wm. Smith first to work out principle
Determination of equivalence in time-stratigraphic age in 2 areas
Correlation has two main tasks:
1. determine relative ages in local area
2. relate local ages to standard geologic time scale
Physically match by grain size, color, sed. structures, sequences
Key bed, widespread, thin, distinctive units, *i.e.* ash falls
Index fossil, common, widespread, restricted geologic age
Assemblages more useful than a single species

## The Geologic Column and the Geologic Time Scale

Geologic column, composite of strata without gaps or overlaps
Subdivisions: eons, eras, periods, epochs

*Eons*

Largest intervals of geologic time, 4 of them

*Eras*

Major subdivisions of eons, based upon life-form content

*Periods*

Subdivisions of eras of the Phanerozoic Eon, geographical/litholo

*Epochs*

Subdivisions of periods, especially Tertiary, % of extant animals
The names of the above units have standard worldwide names

# Measuring Absolute Geologic Time

Absolute time has always haunted geologists and mankind

*Indirect Attempts to Measure Absolute Geologic Time*

Thickness of sed. rock column divided by rate of sedimentation

Amount of salt in oceans divided by amount of salt added/year

Rate of heat loss(Lord Kelvin)(talk about concept of assumptions)

1896-discovery of radioactivity was the answer to dilemma

*Radioactivity and the Measurement of Absolute Time*

Isotope, varying atomic mass number because of neutrons

Most isotopes are equally stable, a few are unstable

Unstable parent isotopes disintegrate into stable daughters

Radioactivity involves only the nucleus, not electrons

Three types of radioactive decay:

1. $\beta$ particle emission

2. $\beta$ particle capture

3. $\alpha$ particle emission, also $\gamma$ ray emission

*Rates of Decay*

Rates of radioactive decay are uneffected by geological processes

A constant proportion of parent atoms decay per unit of time

Half-life is the time needed for 1/2 of parent atoms to decay

The sum of parent atoms and daughter atoms is constant

*Potassium-Argon ($^{40}K/^{40}Ar$) Dating*

$^{40}K$ is radioactive with half-life of 1.3 billion years

$^{40}K$ has 2 decay processes:

12% -- $^{40}K + \beta \rightarrow {}^{40}Ar$ , Beta particle capture

88% -- $^{40}K \rightarrow {}^{40}Ca + \beta$ , Beta particle loss

$^{40}K$ is frozen in original crystal structure, $^{40}Ca$ & $^{40}Ar$ remain

$^{40}Ar$ is measured because it is chemically inert

Not $^{40}Ar$ is in initial crystal lattice because the heat drives it out

***Go through age calculation process in text***

*Flexibility of $^{40}K/^{40}Ar$ Dating*

$^{40}K/^{40}Ar$ dating works with very minor quantities of the isotopes

$^{40}K/^{40}Ar$ dating is most successful with pyroclastics & volcanics

can be used on volcanics as young as 50,000 years

*Other Radiometric Dating Methods*

Six commonly used isotopes - $^{238}U$, $^{235}U$, $^{232}Th$, $^{40}K$, $^{87}Rb$, $^{14}C$

*Radiocarbon Dating*

$^{14}C$ is unique: 1 - very short 1/2-life, 2 - daughter not measured

Proportion of $^{14}C$ is nearly constant in atmosphere

Proportion of $^{14}C$ is nearly constant in living organisms

$^{14}C$ is especially useful in archeology and Pleistocene geology

*Absolute Time and the Geologic Time Scale*

Position of igneous rocks will age date the associated sediments

Age dating sediments will age date the contained fossils

***See Fig. 6.16 for technique and go through with class***

20th Cent. geologists have put years onto the 19th Cent. column

The 19th Century geologic column fits absolute ages exactly

Precambrian solid rocks spans 3.9 b.y. to 0.6 b.y., or 3.3 b.y.

Very oldest age date, single mineral grains, is 4.1 b.y.

"Moon dust" and meteorites give ages of 4.6 b.y. to solar system

# The Magnetic Polarity Time Scale

Some rocks contain oriented magnetic minerals

*Magnetism in Rocks*

In magnetic minerals, the atomic magnets line up in parallel

Below the Curie point mineral magnetism orients and is parallel

In frozen basalt, the magnetite orients parallel to Earth's field

Sedimentary rocks may have weak, permanent magnetism

*The Polarity - Reversal Time Scale*

Some rocks(lavas particularly) have reverse polarity

Combination of radiometric age dating and polarity reversals

Fig. 6.20 polarity record for last 20 m.y.

Identifying which magnetic you have is the most difficult part

If an approximate fossil age exists, exact reversal age is "simple"

# Dating Prehistoric Events

Hawaiian basalts interbedded with Pleistocene volcanics

Ethiopian hominids dated by combo. of magnetics and radiomet.

# OVERVIEW

Geologic time is always one of the topics that is of interest to students. Alas, it is a topic that is often difficult to comprehend. The difference between relative and absolute time often demands careful attention on the part of the lecturer.

Relative time can be demonstrated by going through a students day and simply putting the events in order and asking what happens if events get out of order(putting socks on after shoes).

Absolute time can be demonstrated in a similar manner except you put clock times on each activity and you can calculate the time lapse if you get the difference between two events.

The principles of relative time lend themselves well to slides or chalkboard drawings. Absolute time requires more mental gymnastics on the part of the average student. Use any sort of visual aids possible( if only a piece of chalk to represent a sample of Uranium). Do not bog yourself down with a lot of calculations unless you have an exceptional group of mathematically inclined students.

# LEARNING OBJECTIVES

1. Know that strata provide the basis for reconstructing Earth history and past events.

2. Know that most strata are deposited in a nearly horizontal attitude (Original Horizontality).

3. Know that strata are deposited in sequence from oldest on the bottom to youngest on top (Stratigraphic Superposition).

4. Know that stratigraphic superposition concerns relative time and that the ages of two strata can be determined according to whether one of the layers lies above or below the other.

5. Know that unconformities are physical breaks in a stratigraphic sequence marking a period of time when sedimentation ceased and erosion removed some of the  previously laid strata.

6. Know what the three types of unconformities are and be able to distinguish each.

7. Know that a formation is a fundamental rock unit for field mapping distinguished on the basis of its distinctive physical characteristics and that it is names for a geographic locality.

8. Know that systems are rock sequences that accumulated during a specific time interval and that they are the fundamental time-stratigraphic units used to construct the geologic column.

9. Know that geologic time units are based on time-stratigraphic units and represent the time intervals during which the corresponding systems accumulated.

10. know that correlation of strata from place to place is based on physical and biological criteria that permit demonstration of time equivalence and that they are most reliable if several criteria are used.

11. Know that the geologic column is a composite section of all known strata, arranged on the basis of their contained fossils or other age criteria.

12. Know that the geologic time scale is a hierarchy of time units established on the basis of corresponding time-stratigraphic units.

13. Know that systems(time-stratigraphic units) and periods(geologic-time units) are based on type sections or type areas in Europe and North America.

14. Know that the geologic time scale is the global standard to which geologists correlate local sequences of strata.

15. Know that decay of radioactive isotopes of various chemical elements is the basis of radiometric dating.

16. Know the main radioactive isotopes and their daughters: $^{238}U/^{206}Pb$, $^{235}U/^{207}Pb$, $^{232}Th/^{208}Pb$, $^{87}Rb/^{87}Sr$, $^{40}K/^{40}Ar$, $^{14}C/^{12}C$.

17. Know that a sedimentary rock layer can be dated radiometrically when it is bracketed between two bodies of igneous rocks to which a radiometric dating method can be applied.

18. Know that radiocarbon dating is only effective for relatively young organic materials(less than 70,00 years).

19. Know the age of the Earth, determined by uranium-lead dating is 4.6 billion years.

20. Know that magnetism in rocks and the polarity-reversal time scale are useful for dating oceanic crust, lavas, and young sedimentary rocks.

## Lecture Demonstrations:

1. Many of the relative age dating principles can be demonstrated with a layer cake. The layers can be iced, stacked, sprinkles, jelly, etc dropped on them, intrusions between the layers, cross cutting, erosion showing the

"basement complex" and finally total erosion( you and the class eat the cake. A little imagination can also create various unconformable relationships.

2. This is a good lecture for lots of slides to demonstrate the topics.

3. Half-life can be demonstrated with the cake(above) by cutting it in half, then that half on half(1/4's), and so on and on.

## FILMS AND VIDEO CASSETTES

1. Reflections on Time. 23 min., 1969,(source: EB, BU, IU, USC).

2. Time Is. 30 min., 1964, (source: MGH, IU).

3. History Layer by Layer. 23 min., 1967, (source: MGH, IU, USC).

4. Long Time Intervals. 24 min., 1959, (source: MLA, IU).

5. Measuring Geologic Time. 27 min., 1960, (source: UT).

6. Geologic Time.24 min., 1986 (source: EB).

7. Eighteen Corridors of Time.60 min., 1983, (sources: BBC).

8. Radioactive Dating. 13 min., 1981, (source: CFV).

9. Fossils. Exploring the Past. 16 min., 1978, (source: EB).

10. Earth Science, Exploring Planet Earth. 20 min. 1978, (source: EB).

11. Message in the Rocks. 57 min., 1982, (source: TLV).

12. Five Billion Years. 7 min., (source: MGH).

## REFERENCES

Berry, W. B. N., 1968, Growth of the Prehistoric Time Scale. San Francisco: Freeman.

Chamberlain, T. C., 1965, The Method of Multiple Working Hypotheses, Science. v. 148, pp. 754-759.

Cloud, R., 1977, Scientific Creationism-A New Inquisition Brewing, The Humanist. v. 37, pp. 6 16.

Eicher, D. L., 1968, <u>Geologic Time.</u> Englewood Cliffs, NJ: Prentice-Hall.

Faul, H., 1966, <u>Ages of Rocks, Planets and Stars.</u> New York: McGraw-Hill.

Gould, S. J., 1965, Is Uniformitarianist Necessary?, <u>American Journal of Science.</u> v. 263, pp. 223-228.

Haber, F. C., 1959, <u>The Age of the World: Moses to Darwin.</u> Baltimore: Johns Hopkins Press.

Hamilton, E. I., and R. M. Farquhar, 1968, <u>Radiometric dating for Geologists.</u> New York: Wiley.

Harland, W. B., A. V. Cox, P. G. Llewellyn, C. A. G. Pickton, A. G. Pickton, and R. Walter, 1982, <u>A Geologic Time Scale.</u> London: Cambridge Univ. Press.

Hume, J. D., 1978, An Understanding of Geological Time, <u>Journal of Geological Education.</u> V. 26, pp. 141-143.

Palmer, A. R., 1983, The Decade of North American Geology 1983 Geologic Time Scale, <u>Geology.</u> v. 11, pp. 503-504.

Toulmin, S., and J. Goodfield, 1965, <u>The Discovery of Time.</u> New York: Harper and Row.

# CHAPTER 7

## WEATHERING AND SOILS

***Vignette on damage to carbonate rocks by pollutants***

    Carbonate building stone, limestone and marble, decays easily

    $SO_2$ and acid rain dissolve limestone and marble structures

## Weathering

    Wherever air and water are present, minerals are altered

    Weathering - chemical alteration and mechanical breakdown

    Weathering progresses from surface downward

    Chemical weathering change minerals into new stable ones

    Mechanical weathering breaks rocks into smaller pieces

*Mechanical Weathering*

*Development of Joints*

    Joints develop within 50m of surface as erosion removes crust

    Joints act as passageways for rainwater and more weathering

    Colunmar joints, shrinkage features in sheetlike igneous bodies

*Crystal Growth*

    Deserts or near springs as evaporation causes crystal growth

    May occur in polar climates with very minor amts. of water

*Frost Wedging*

    Very effective where many freeze-thaw cycles occur annually

    Freezing water in cracks pries rocks apart

*Effects of Heat*

    Diurnal expansion-contraction due to heating, still in debate

    Fire is a very effective mechanical weathering agent

*Plant Roots*

    Roots from small mosses to large trees are wedging agents

*Chemical Weathering*

    Exposed high temperature/pressure Ig-Met minerals, unstable

    High temperature/high precipitation promote chem weathering

*Chemical Weathering of Rock-Forming Minerals*

Rainwater dissolves $CO_2$ to produce carbonic acid

*Hydrolysis*

When $H^{1+}$ and $OH^{1-}$ from water replace ions in minerals

An important chemical weathering process

*Leaching*

Continued removal of soluble matter by water solutions

*Oxidation*

The loss of electron(s) by ions *i.e.*, $Fe^{2+}$ to $Fe^{3+}$

Hydration commonly associated, addition of water to lattice

*Effects of Chemical Weathering on Common Rocks*

***Go over Table 7.1***

The most common weathering products are clay plus other ions

Output varies with parent rock being weathered

*Concentration of Stable Minerals*

Quartz, gold, platinum, diamond persist through weathering

*Weathering Rinds*

Most rocks develop a rind as weathering progresses inward

Thickness of weathering rinds suggests relative time of activity

*Exfoliation and Spheroidal Weathering*

Exfoliation, process of thin layers of weathered material spalling

As exfoliation progresses, blocks become more spherical

Chemical weathering generally produces a volume increase

Spheroidal weathering is subsurface along intersecting joints

Effectiveness of chemical weathering is proportional to surface

Surface area increases as mechanical weathering progresses

*Factors That Influence Weathering*

*Rock Type and Structure*

Rocks with a high quartz content resist chemical weathering

Rocks with a high quartz content tend to produce hilly terrain

Massive rocks weather more slowly than intensely jointed ones

Differential weathering, contrast of rocks types, local topography

Resistant rocks produce cliffs, nonresistant rocks produce slope

*Slope*

Steep slopes are fresher rock than gentle slopes

*Climate*

    High temperature-moisture increases weathering rates

    Low temperature-moisture decreases weathering rates

    Carbonates produce cliffs in deserts, valleys in hot-humid

*Burrowing Animals*

    Move material so it is more easily weathered

    Great variety - worms, insects, mammals, etc.

*Time*

    Intensity of weathering vary as mineralogy/climate during time

    Weathering rates decrease with time toward a steady-state

## Soils

    Basis of stability of terrestrial biosphere and hydrologic cycle

*Origin*

    Chem-Mech breakdown of rock combined with organic activity

    Organic matter is derived from decay of plants and animals

*Soil Profile*

    Soil horizons evolve as weathering progresses--Fig. 7.17

    O horizon, uppermost organic matter rich layer

    A horizon, dark due to humus with minerals, zone of loss

    E horizon, light due to lack of dark oxides, evergreen forests

    B horizon, brown to red, enriched in clay and Fe and Al oxides

    K horizon, desert horizon of calcium carbonate accumulation

    C horizon, weathering parent material

*Soil Types*

    Six soil-forming factors

    Soils classified according to their physical-chemical properties

*Polar Soils*

    Entisols, Histosols, Inceptisols - generally poorly developed

*Temperate-Latitude Soils*

    Alfisols, Spodisols, Entisols, Inceptisols, Mollisols, Ultisols

*Desert Soils*

    Aridisols - lack of water slows rate and intensity of development

*Tropical Soils*

> Oxisols, Vertisols - extreme chemical alteration

> Extreme weathering causes high iron oxide material to develop

*Rate of Soil Formation*

> Soil profiles develop more rapidly than chemical weathering

> Soils develop in an orderly manner due to weathering, plants

> Less moisture the slower soil formation progresses

*Paleosols*

> Soil formed at surface and subsequently buried

> Important clues to former landscapes, vegetation cover, climate

*Soil Erosion*

> Population increases, less desirable ag land, higher erosion rates

> Loss of natural vegetation, destabilization and fertility losses

*Indirect Effects of Soil Erosion*

> Filling of reservoirs, burial of ag land, killing of wild life

*Rates of Soil Loss*

> Erosion removes topsoil at rates of 5+ times its rate of formation

> Approximately 6.7 kg of soils is lost for each 1 kg food produced

*Control of Soil Erosion*

> Effective controls exist, but must be used extensively/intensively

> As much land as possible should be in grass, solid or strips

> Crop rotation of row crops and solid cover crops

> Steeps slopes should not be farmed, or else terraced

*Soil Erosion and the World Economy*

> On a short term basis (decades) soils are not renewable

> Agriculture is foundation of world economy

## Mineral Deposits Formed by Weathering

> More soluble are removed and less soluble ones concentrated

*Lateritic Concentrations*

> Bauxite is the most valuable of the group

*Secondary Enrichment*

> Removal of silica in solution well leave ore minerals behind

> Weathering sulfide minerals produce sulfuric acid that reacts

## Weathering and the Rock Cycle

        Plate tectonics constantly brings new material to the surface

        High temp/pressure minerals weather most readily

        Mineral stability is reverse of Bowen's Reaction Series, Tab. 7.3

## Plate Tectonics and Global Weathering Rates

        High-altitude land masses are directly related to weathering rate

            1. Rapid uplift, high mechanical weathering, rapid erosion

            2. Most dissolved mineral matter from sedimentary terrain

            3. High mountains receive high amounts of precipitation

        Present weathering rates are among highest is history of Earth

        Rates of mountain uplift parallel rates of seafloor spreading

## Overview

The processes of weathering and soil formation are among the most critical to the well being of mankind.  Sadly these processes are not too exciting for beginning geology students.  You will need to use all of the imagination you have to present stories and illustrations that your particular students will relate to.  These lectures go better if you can excite them even a little at the out set.

Mechanical weathering is always easiest to demonstrate or call examples to mind.  A couple like unloading and exfoliation are difficult to visualize, slides help.  Chemical weathering is more difficult to deal with because they are chemistry and most students have not had it, or did not learn it or both.  If you do reactions in symbols be sure to spend some more time explaining them.  I have found it as effective to do the reactions in words and where possible have samples of the mineral, water, etc. going through the reaction.

Soil formation is the interrelation of climate, organisms, topography, parent material, and time.  Slides helps understand these factors and the final product.  Any local feature that can be brought into these lectures helps greatly.

## LEARNING OBJECTIVES

1. Know that weathering extends to whatever depth air and water penetrate the Earth's crust.

2. Know that water solutions, which enter the bedrock along joints and other openings, attack the rock chemically and physically, causing breakdown and decay.

3. Mechanical and chemical weathering, although involving very different processes, generally work together.

4. Know the several mechanical weathering processes.

5. know that chemical weathering involves the transformation of minerals formed at high temperature and/or high pressures into minerals that are stable at the Earth's surface.

6. Know the several processes of chemical weathering.

7. Know that breaking large rocks into many smaller pieces increases the surface area and accelerates the rate of chemical weathering.

8. Know that the effectiveness of weathering depends on rock type, structure, slope of the land, local climate and the time the processes have to operate.

9. Know that heat and moisture speed up chemical reactions and that chemical weathering is much more active in warm, moist climates than in cold, dry climates.

10. Know that soils consist of weathered regolith and are capable of supporting plants.

11. know that soils develop distinctive horizons that are the result of the interactions of climate, vegetation cover, soil organisms, parent material, topography, and time.

12. Know the characteristics of the soil horizons and how to distinguish each.

13 Know that the soil classification used in the United states places soils in ten orders based upon their physical characteristics.

14. Know that paleosols are buried soil and that they provide clues to former landscapes, plant cover, and climate.

15. Know that paleosols are useful for subdividing, correlating, and dating the strata in which they are found.

16. Know that soil erosion and degradation are global problems that have been increasing as world population rises.

17. Know that effective control measures include crop rotation, terracing, and tree planting, but that halting widespread loss of soils is a formidable challenge.

18. Know that chemical weathering causes concentration of economically valuable mineral deposits that are primary sources of aluminium nickle, iron, manganese, and copper.

19. Know that weathering is an integral part of the rock cycle.

20. Know that minerals least resistant to weathering are those that form in crustal environments most different from those at the Earth's surface.

21. Know that global rates of chemical weathering are linked to the presence or absence of high mountains, and thus to plate tectonics.

## LECTURE DEMONSTRATIONS

1. The force of freezing water can be easily demonstrated with a 6 - 8 inch piece of iron or plastic pipe filled with water and with a cap securely attached to each end. If liquid air is available this can be done in one class period.

2. color slides of weathering features and soils are one of the principal hope to make these lectures exciting and hold the students attention.

3. Samples showing weathering help a great deal.

4. A soil monolith helps in both lecture and laboratory. See Carver(1971, chap. 10).

## FILMS AND VIDEO CASSETTES

1. Erosion: Leveling the Land, 13 min., 1964, (source: EB, BU, IU, USC).

2. Erosion and Weathering: Looking at the Land, 17 min., 1976, (source: EB).

3. Evolution of Landscapes, 19 min., 1986, (source: EB).

4. Rock Weathering - Origin of Soils, 26 min., 1960, ( source: UT).

5. Uneventful Day, 29 min., (source: FI).

## REFERENCES

Carrol, D., 1970, Rock Weathering, New York: Plenum Press.

Carver, R. E. (ed.), 1971, Procedures in Sedimentary Petrology, New York: Wiley-Interscience.

Fitzpatrick, E. A., 1971, Pedology - A Systematic Approach to Soil Science, Edinburgh: Oliver and Boyd.

Garrels, R. M., and F. T. Mackenzie, 1971, Evolution of Sedimentary Rocks, New York: Norton.

Goldich, S. S., 1938, A Study of Rock Weathering, Journal of Geology, v. 46, pp. 17-58.

Knapp, B., 1979, Soil Processes, Boston: Allen and Unwin.

Legget, R. F., 1967, Soil: Its geology and Use, Geological Society of America Bulletin, v. 78, pp. 1433-1459.

Loughnan, F. C., 1969, Chemical Weathering of the Silicate Minerals, New York: Elsevier.

Ollier, C. D., 1969, Weathering, New York: Elsevier.

Small, R. J., and M. J. Clark, 1982, Slopes and Weathering, New York: Cambridge Press.

Twidale, C. R., 1973, On the Origin of Sheet Jointing, Rock Mechanics, v. 5, pp. 163-187.

# CHAPTER 8

## MASS-WASTING

### OUTLINE

#### *Vignette on relationship of man to volcanic peaks*

Great volcanos of Andes with dense population at base

Nevado del Ruíz, Colombia, 20,000 deaths in 1984, mass-waste

#### **Mass Movement and Slopes**

Mass-wasting, movement of regolith under pull of gravity

Steady-state, input of material on a slope equals output at base

Mass-wasting is both subaerial and subaqueous

#### **Role of Gravity**

Rock particles move down slope

Shear stress tends to move particles downslope

Shear strength, internal resistance to downhill movement

Stability when shear strength > shear stress

The steeper the slope the greater the shear stress

Mass-wasting is common in mountains, result of plate tectonics

#### **Role of Water**

Small amounts of water increase strength of sediments

Too much water decreases strength of sediments

Damp clay - silt is strong, wet it will flow

Much water along a rock contact may cause failure by buoyancy

Water reduces shear strength by:

1. reduces natural cohesiveness between grains

2. reduces friction at base by increased fluid pressure

## Mass-Wasting Processes

Mass-wasting takes place on slopes, but many processes

This text classification:

1. Failure results in downslope transfer of coherent masses

2. Downslope movement of sediment, water, air mixtures

*Slope Failure*

After failure, rock debris is transferred downhill to new stability

*Slumps*

Downward and outward rotation of curved concave-up surface

Common along oversteepened roads and stream banks/seacoasts

Commonly associated with heavy rains or sudden shocks

## Falls and Slides

Rockfall, free falling detached rock bodies of bedrock from cliffs

May be a single piece or very large masses of rock

Debris fall, free falling detached rocks with regolith and plants

Rockslide, detached mass of rock moving down a sloping surface

Debris slide, similar to a rock slide, but with vegetation, regolith

Talus, the accumulation of loose material at foot of a cliff

Angle of repose, steepest surface angle of loose material

*Sediment Flows*

Flow of mixtures of rock, water, and air

*Factors Controlling Flow*

The way sediment flows depends on:

1. relative proportions of solid, water, and air

2. physical/chemical properties of the sediment

Continuum of flows from clear stream water to stationary masses

Types of sediment flows:

1. slurry flow, a moving mass of water - saturated sediment

2. granular flow, sediment - water - air mix, grain supported

Further subdivided by rate of flow

*Slurry Flows*

Dense sediment mixture, results are nonsorted sediment

*Solifluction*

Very slow(30 cm/yr), saturated soil - regolith, lobes and sheets

*Debris Flows*

    Unconsolidated regolith, mostly > sand, 1 m/yr to 100 km/hr

    Move in channels and end in a tonguelike front, irregular surface

*Mudflows*

    Water content great enough to be highly fluid, > 1 km/ hr

    Running water, get more mud, dams, breaks into open as sheets

    Active volcanos are notorious for mudflows during eruptions

*Granular Flows*

    Air filled pores or water easily released

*Creep and Colluvium*

    Creep is imperceptibly slow downslope movement of regolith

    Steeper the slope the greater the creep rate

    The greater the vegetation cover the slower the rate of creep

    Colluvium - loose, incoherent creep deposits, unsorted, angular

*Earthflows*

    Common, granular, 1 m/day to x00 m/ hr, < sand size, long lived

    Typically head in a steep scarp, long, narrow tonguelike

    Movement associated with water pressure along shear surfaces

    Liquifaction occurs when saturated material is shocked

*Grain Flows*

    Movement of dry, granular material at > the angle of repose

*Debris Avalanches*

    Extremely large granular flows, ten to hundreds of km/hr

    These have been some of the most disastrous events to humans

    Infrequent and difficult to study

    Flanks of large stratovolcanos are particularly susceptible

*Mass-Wasting in Cold-Climates*

    Mass-wasting is especially active in high latitudes and altitudes

*Frost Heaving and Creep*

    As ice freezes, the regolith mass is pushed up

    Frost heaving strongly influences creep in freeze - thaw cycles

*Gelifluction*

    The process of a thin, thawed layer over a frozen layer moves

*Rock Glaciers*

    A moving tongue or lobe of ice-cemented rock debris

*Subaqueous Mass-Wasting*

Extremely common and widespread on seafloor

Produce turbidity currents, source of significant amounts of sed.

*Marine Deltas*

Common on delta, source zone, central channel, deposition zone

*Mass-Wasting in the Western North Atlantic*

Vast areas covered by large slides, 40,000 km$^2$ by 50 m thick

Submarine canyons may have formed by slumping activities

Generally move on slopes > 20°, but less if shocked

Most mass-wasting occurred during the last ice age

*Hawaiian Submarine Landslides*

Numerous chaotic masses, scarps, anomalous corals suggest this

## Triggering of Mass-Wasting Events

Sometimes seen random, but many are related to an occurrence

*Shocks*

Shocks, *i.e.*, earthquakes, may release energy to trigger failures

*Slope Modification*

Oversteepening of slopes by human activity sets the stage

*Undercutting*

Stream or surf action will oversteepen banks and then collapse

*Exceptional Precipitation*

Excessive amounts of water add weight and buoyancy to grains

*Volcanic Eruptions*

Stratovolcanos, ejecta, loose unstable debris, water, ice melting

*Submarine Slope Failures*

Numerous factors lead to submarine slope failures

## Mass-Wasting Hazards

Mass-wasting affects more people as the population expands

*Assessments of Hazards*

Frequency curves are based upon past events

Maps of areas impacted in the past can be constructed

*Mitigation of Hazards*

Many hazards can be mitigated by serious advanced planning

## Himalayan Landslides and Plate Tectonics

Most historic and prehistoric landslides are at converging plates

1. World's highest mountains are at or near plate boundaries

2. Most large earthquakes are at or near plate boundaries

Rapid uplift rates means rapid erosion and mass-wasting rates

## OVERVIEW

Mass wasting is the group of processes that are responsible for down slope movement of weathered material under the influence of gravity. Mass wasting processes are intermediate between weathering and streams that are going to transport the material further. Distinguish between weathering, mass-wasting, and erosion. these processes that go on simultaneously are easy to confuse.

Many of the mass-wasting processes are not very spectacular, but effect the lives of the students constantly. Use your experience in local mass-wasting to illustrate your lectures. Examples are to be found virtually everywhere and they will be much more interesting than the ones from a different region. Slides are of great value to this lecture topic because most students have little experience with the topic.

There are some classic disasters associated with this topic and at least some of your students will be acquainted with them.

## LEARNING OBJECTIVES

1. Know that mass-wasting causes rock debris to move downslope under the influence of gravity without a carrying agent.

2. Know that mass-wasting occurs both on land and beneath the oceans.

3. Know that the composition and texture of debris, the amount of air and water mixed with it, and the steepness of slope influence the type and velocity of slope movements.

4. Know that mass-wasting processes include sudden slope failures and downslope flow of mixtures of sediment, water, and air.

5. Know that failures occur when the shear stress reaches or exceeds the strength of slope materials.

6. Know that High water pressure in rock voids or sediment reduces shear strength and increases the likelihood of failure.

7. Know that slumps involve a rotational movement along a concave-up surface that results in backward-tilted blocks of rock or regolith.

8. Know that falling and sliding masses of rock and debris are common in mountains where steep slopes abound.

9. Know that rockfall debris accumulates at the base of a cliff to produce a talus with slopes that stand at the angle of repose.

10. Know that slurry flows involve dense moving masses of water-saturated sediment that form nonsorted deposits when flow ceases and that flow rates range from very slow to fast.

11. Know that granular flow is dry or water escapes quickly allowing sediment to be in constant grain-to-grain contact.

12. Know that creep is an important downslope process even though it is slow it is so widespread.

13. Know that large, rapidly moving debris avalanches are relatively infrequent but potentially hazardous to humans.

14. Know that in regions of perennially frozen ground, frost heaving, creep, and gelifluction are quantitatively important mass-wasting processes.

15. Know that large areas of seafloor on the continental slopes show evidence of widespread slumps, slides, and flows.

16. Know that submarine mass-wasting was especially active during glacial ages when sea level was lower and large quantities of stream sediment were transported to the edge of the continental shelves.

17. Know that slope failures can be triggered by earthquakes, undercutting by streams, heavy or prolonged rains, or volcanic eruptions.

18. Know that subaqueous slope failures are frequently related to rapid deposition of sediments, oversteepening of slopes, and earthquake shocks.

19. Know that the loss of life and property from mass-wasting events can be prevented or mitigated by advanced assessment and planning based on geologic studies of previous occurrences.

## LECTURE DEMONSTRATIONS

1. An interesting demonstration of water diminishing resistance to flow is by placing a can with hole punched in the bottom on an inclined piece of glass. by pouring water into the can it will readily slide down the glass.

2. Slope failures can be demonstrated by placing a block of rock on top of a pile of sand, then cutting away sand around the base. A similar experiment can be done, only add water to the pile of sand and watch it flow. This can

get messy, go outside or do it in an aquarium, or videotape it and show the class.

3. Stimson(1980) illustrates a slops failure device.

4. GeoPhoto Publishing has available slide sets of the Thistle, Utah, landslide, the Mount Shasta, California, debris avalanche and the 1978 Laguna landslide in California.

## FILMS AND VIDEO CASSETTES

1. Erosion - Leveling the Land, 14 min., 1964, (source: EB).

2. Downslope Movement, 27 min., 1960, (source: UT).

## REFERENCES

Bromhead, E. N., 1986, The Stability of Slopes, New York: Chapman and Hall.

Coats, D. R., 1977, Landslides, Geological Society of America Reviews in Engineering Geology, v. III.

Coates, D. R., 1981, Environmental Geology, New York: Wiley.

Crandall, D. R., 1971, Postglacial Lahars from Mount Rainier Volcano, Washington,U,S. Geological Survey Professional Paper 677.

Kiersch, G. A., 1964, Vaiont Reservoir Disaster, Civil Engineering, v. 34, pp.32-39.

Leveson, D., 1980, Geology and the Urban Environment, New York: Oxford Univ. Press.

Lowe, D. R., 1986, Lahars Initiated by the 13 November 1985 Eruption of Nevado del Ruiz, Colombia, Nature, v. 324, 51-53.

Selby, M. J., 1982, Hillslope Materials and Processes, New York: Oxford Univ. Press.

Small, R. J., and M. J. Clark, 1982, Slopes and Weathering, New York: Cambridge Univ. Press.

Stimson, B., 1980, Inexpensive device for demonstrating Rock Slope Failure and Other Collapse Phenomena, Journal of geological Education,v. 29, pp. 103-104.

Voight, B., 1978, Rockslides and Avalanches, part I, Natural Phenomena, New York: Elsevier.

# CHAPTER 9

## STREAMS AND DRAINAGE SYSTEMS

### OUTLINE

**Vignette on channel changes of the Mississippi River**

Excerpt from "Life on the Mississippi" by Mark Twain

### Streams in the Landscape

Nearly the whole land surface shows evidence of stream erosion

Stream - body of water flowing down hill in a natural passageway

Channel - natural passageway of a stream

Load - the detritus a stream moves

Discharge - quantity of water passing a point per unit of time

Streams play many complicated, interrelated roles

1. Avenue of transportation

2. Building sites

3. Agricultural resource

4. Transport water, detritus, solutes

5. Sculpture the land surface

### Stream Channels

Conduit for running water, discharge changes in time and space

*Cross-Section Shape*

Small streams are narrow and deep, large are wide and shallow

Shape changes over the length of the stream

*Long Profile*

Gradient is the amount of drop per horizontal travel unit

Gradient decreases down stream, but not smoothly

# Dynamics of Streamflow

Runoff - portion of precipitation that flows over the land surface

Overland flow - the initial surface water before it is channelized

Streamflow - flowing water that is channelized

## Factors in Streamflow

Factors that control stream behavior

1. Gradient
2. Channel cross-section area
3. Average velocity
4. Discharge
5. Load

Discharge = Cross-section area of channel X Average velocity

During flooding water level goes up and bottom goes down

At any moment in time the channel-stream-discharge balance

## Changes Downstream

Traveling from head to mouth along a stream

1. Discharge increases
2. Channel cross-sectional area increases
3. Velocity increases slightly
4. Gradient decreases

## Floods

When discharge exceeds channel capacity and overflow occurs

A storm will produce a major peak on a hydrograph

As discharge increases, velocity increases, increasing work level

## Flood Prediction

Past floods are plotted on flood-frequency curves

Flood-frequency curves give recurrence interval of any size flood

Flood-frequency curves assume the future will be same as past

## Catastrophic Floods

Occur only rarely in geologic history

***Talk about J Harlen Bretz and the Channeled Scablands***

## Base Level

Limiting level below which a stream cannot erode the land

Generally sealevel, remember global sealevel fluctuates

*Natural and Artificial Dams*

    Landslide deposit, glacial/stream deposits, lava flows, glacier ice

    Artificial dams are temporary, trap sediment, disrupt flow

*Hydroelectric Power*

    Produced as gravity drops water through turbines

    Potential recoverable energy is great, actual is much smaller

## Channel Patterns

    Variety of stream size and shape is result of gradient, discharge

*Straight Channels*

    Only for short distances

    Water migrates back and forth producing bars on alternate sides

*Meandering Channels*

    A series of similar sized smooth bends

    Velocity is lowest at banks and bottom, max at midchannel top

    Meander migrate slowly downstream

    Point bar - accumulation of sediment inside a meander

    Is bank sediment is uniform, meanders form uniform size/rate

    Meanders get cutoff and and are left as an oxbow or oxbow lake

    As meanders are cutoff, new ones form

*Braided Channels*

    Water repeatedly merges then splits as deposition/erosion occur

    Common with highly variable discharge and easily eroded banks

    Even though a braided stream is wide, only 10% has water

## Erosion by Running Water

    Sheet erosion - combination of raindrop impact & overland flow

    Raindrop impact - raindrops hit bare ground dislodging clasts

    Overland flow - runoff before a stream channelizes

    Vegetated areas less prone to sheet erosion

    Laminar flow - thin layer along channel, parallel molecule move.

    Turbulent flow - erratic, complex, swirls, eddies

    Turbulence and velocity determine ability to erode and transport

# The Stream's Load

Load is 2 parts - bed load and suspended load

Alluvium - detrital sediment deposited by a stream

Dissolved load is chiefly product of chemical weathering

*Bed Load*

Bed load - rolling, sliding, saltation

*Placer Deposits*

Heavy minerals, *i.e.* gold, is deposited with coarser, lighter mins.

Placer - a deposit of heavy minerals concentrated mechanically

Placer minerals -gold, platinum, copper, tinstone, diamond, ruby

Steps - mechanical/chemical weathering, mass-wasting, concent

*Suspended Load*

Clay and silt mechanically carried by a stream

Carried due to upward directed portion of stream turbulence

*Dissolved Load*

Chemical weathering products - dominated by 7 ions

Streams with major groundwater contributions have higher load

*Downstream Changes in Grain Size*

Generally, grain size decreases downstream

Result of sorting and abrasion

*Downstream Changes in Composition*

Load changes as different sources and tributaries merge

*Sediment Yield*

Amount of sediment eroded by runoff & transported by streams

Related to geologic, climatic, topographic factors

Highest sed. yield is from tropics with high chemical weathering

Temperate climates with dense vegetative cover are minimum

Extreme yields from steep mountains at plate boundaries

Humans generally increase sediment yield

# Stream Deposits

Deposition occurs if a stream loses gradient, velocity, discharge

*Floodplains and Levees*

Floodplain - flat portion of valley covered by flood waters

Natural levee - broad, low ridge of fine alluvium along channel

*Terraces*

    Remnant of an old floodplain that has been cut into by channel

    Downcutting result of changes in discharge, load, gradient, base

*Alluvial Fans*

    Deposit of detritus in a flat valley at mouth of steep canyon

    Deposition occurs where infiltration causes discharge loss

    Exact shape of fan is determined by load, discharge, infiltration

*Deltas*

    Formed where a stream enters a standing water body & deposits

    Fan delta - coarse grained, alluvial fan builds into water body

    Braid delta - coarse grained, braided stream into water body

    Bed load dropped coarsest to finest, suspended coarsest to fine

    Delta facies distinct - foreset, bottomset, topset

    Many large rivers have massive deltas, each peculiar from others

    Fine portion of suspended load settles on sloping delta front

    Distributaries, currents, and loads determine the final delta form

## Drainage Systems

    Streams are organized into intricate drainage systems

*Drainage Basins and Divides*

    Drainage basin - the total area that contributes water to a stream

    Divide - boundary of a drainage basin

*Stream Order*

    First order - smallest stream with no tributaries

    Second order - merging of 2 first order streams etc.

    ***Go over this concept carefully, easily confused***

    There are geometrically more of each smaller order of streams

    Stream systems are just right size & spacing for most efficiency

*Evolution of Drainage*

    Drainages develop very quickly and are highly systematic

    Stream capture - gain and/or loss of tributaries

    Streams move toward hydraulic equilibrium

*Drainage Patterns, Rock Structure, and Stream History*

    Stream pattern are controlled by geologic structure, land history

    Read rock type, direction and amt. of dip, folds, joints, offsets

    See Fig. 9.32

*Continental Divides*

A ridge separating water flow into 2 major oceans

Plate tectonic controls primary divides and drainage basins

Divides may be steep and sharp(Rockies), broad(Arctic-Atlantic)

Divides may last hundreds of millions or tens of thousands years

## Tampering with the Niles

Man lived with Nile and its annual flood for millennia

Great population increases put great demands on available water

Aswan dam constructed in 1960's for water and hydropower

Incredible downstream problems have developed

# OVERVIEW

Streams are probably the most nearly ubiquitous feature on the surface of the planet. They are probably the geological agent most familiar to students. Students have some personal concept of a stream and this can be used to advantage. This is opposed to several concepts that have been dealt with previously that students have no relationship with. This chapter opens a section in the text that students will feel "at home" with.

The idea that streams are dynamic systems is repeated throughout the text and is a good unifying theme for lectures. Open systems are fairly easy to grasp and movement toward equilibrium by stream helps lead students along through the lectures. Dynamics and equilibrium help discuss sculpture of the lands.

Discussion of the measurable variables is straight forward because students have some grasp of measurement. Gradient, discharge, length, order, etc. develop smoothly in most situations. Slides will help once again with the concepts.

Competence and capacity - load are a bit more abstract and may be confused. Thinking up examples and situations of local interest will help.

Floods are always exciting and every place has some sort of a flood record - use local examples whenever possible, they are much more interesting than the flood in someone else's town. This is also a good time to deal with man's interrelationship with streams and the great havoc man and streams can deal to each other. Again local examples!

# LEARNING OBJECTIVES

1. Be able to list the evidence that leads us to think that streams are a major force shaping the Earth's landscape.

2. Know the difference between overland flow and streamflow.

3. Know why vegetation decreases the effectiveness of sheet erosion.

4. Know why a stream's channel dimensions (depth, width) and velocity adjust in response to changes in discharge.

5. Be able to describe the controlling factors that would have to change in order for a braided stream channel to change to a meandering system, and the reverse.

6. Know why stream velocity generally increases downstream, despite a decrease in stream gradient.

7. Know why erosion occur downstream when a dam is built across a channel.

8. Know what is meant by a "200-year flood".

9. Know what the effect of a "200-year flood" is on landscape evolution.

10. Know what factors would have to change in order for a stream to start transporting fine gravel in suspension when it had been bedload.

11. Know the methods of movement of sediment when it is bedload vs. suspended load.

12. Know why average sediment size decreases downstream when average velocity increases downstream.

13. Know how increased urbanization affects erosion and sediment load in a drainage basin.

14. Know how to distinguish between deltas and alluvial fans based upon features preserved in the stratigraphic record.

15. Know a reasonable explanation for the orderly arrangement of tributaries that is typical of most drainage systems.

## LECTURE DEMONSTRATIONS

1. If a flume or stream table is available, a large variety of stream and depositional features can be demonstrated.

2. A large variety of slide sets and individual slides are available from the companies listed elsewhere in this supplement. Also streams are easy to produce your own slide for because they are everywhere in all sizes. The use

of slides cannot be over emphasized.  Also the may films and now video cassettes(these are also easy to produce yourself.)

3. Settling rates are easily demonstrated in a 1000 ml graduated cylinder. Practice a little before going to class and do not wear your best clothes that day.  "Slopping" out of the end is a constant hazard when you are shaking-up the cylinder.

## FILMS AND VIDEO CASSETTES

1. Earth Science - Water versus Land, 16 min., 1976, (source:  IU).

2. Erosion - Leveling the Land, 14 min., 1964, (source: EB)

3. Evolution of Landscapes, 19 min., 1986, (source: EB).

4. Flood Forecasting,20 min., 1986, (source: EB).

5. Flow in Alluvial Channels, 40 min., 1962, (source: USGS).

6. Land Use on the Flood Plains, 16 min., 1976, (source: IU).

7. River Characteristics, 27 min., 1960, (source: UT).

8. Rivers: The Work of Running Water, 22 min., 1981, (source: EBE).

9. The Earth: Action of Rivers, 11 min., 1969, (source: CFV).

10. The Water Crisis, 47 min., 1982, (source: TLV).

## REFERENCES

Baker, V. R., *et al.*,(eds.), 1988, Flood Geomorphology, New York: Wiley.

Bloom, A. L., 1969, The Surface of the Earth, Englewood Cliffs, NJ: Prentice-Hall.

Chorley, R. J. (ed.), 1971, Introduction to Fluvial Processes, New York: Methuen and Co.

Dury, G. H., 1966, The Concept of Grade. In Essays in Geomorphology, ed. G. H. Dury, New York: Elsevier.

Knighton, D., 1984, Fluvial Forms and Processes, London: Edward Arnold.

Leopold, L. B., 1974, Water, a Primer, San Francisco: Freeman.

Leopold, L. B., M. G. Wolman, and J. P. Miller, 1964, Fluvial Processes in Geomorphology, San Francisco: Freeman.

Mayer, L., and D. Nash (eds.), 1987, Catastrophic Flooding, Boston: Allen and Unwin.

Morisawa, M., 1968, Streams: Their Dynamics and Morphology, New York: McGraw-Hill.

Ollier, C., 1981, Tectonics and Landforms, London and New York: Longman.

Petts, G., and I. Foster, 1985, Rivers and Landscape, London: Edward Arnold.

Richards, K., 1982, Rivers, Form and processes in Alluvial Channels, New York: Methuen and Co.

Ward, R., 1978, Floods, a Geographical Perspective, New York: Wiley.

# CHAPTER 10

## GROUNDWATER

### OUTLINE

***Vignette concerning groundwater from alluvial fans in Middle East***

>   Alluvial fans throughout the Middle East have ancient aqueducts
>   Hand dug thousands of years ago to supply water to the valley

### Water in the Ground

>   Most early towns utilized surface water, but soon went to wells
>   Population growth has progressively polluted the water systems

*Origin of Groundwater*

>   Groundwater is 1% of total, 1/3 glaciers, 40x surface fresh water
>   Originates mostly as rainfall, enters ground due to gravity
>   Concept demonstrated by Pierre Perrault for Seine River

*Depth of Groundwater*

>   More than half is w/in 750 m of the surface, considered here

*The Water Table*

>   Zone of aeration - pores mostly filled with air, water occasionally
>   Zone of saturation - pores filled with water
>   Water table - boundary between zones of aeration and saturation
>   Water table is a subdued imitation of land surface in humid clim.
>   Groundwater flows into bottom of streams
>   Water table determines top of usable groundwater resources

### How Groundwater Moves

>   Infiltrates from precipitation and resurfaces in streams and lake
>   Top 750 m water is in very slow rates of motion

*Porosity and Permeability*

        Porosity - percent of void space in rocks

        Porosity effected by size and shape of grains, amount of cement

        Permeability - a measure of how easily a fluid flows through solid

        Permeability determined by pore size, shape, connectedness

        Larger the grains the greater the permeability

*Movement in the Zone of Aeration*

        Upper soil layers retain water on clays for plant use, evapotrans.

        Gravity moves the rest of the water to the zone of aeration

*Movement in the Saturated Zone*

        Percolation - slow movement of water through pores

        Flows down under hills and hydrostatically up under streams

*Recharge and Discharge Areas*

        Recharge - replenishment of groundwater from precipitation

        Recharge area - region of landscape where precipitation seeps in

        Discharge areas - regions where groundwater comes to surface

        Time in ground varies with distance and rate; days to millennia

        Recharge in humid regions is nearly entire landscape

        Recharge in arid regions is mountains and alluvial fans

*Discharge and Velocity*

        Steeper the gradient the faster the rate of flow

        Hydraulic gradient - difference in elevation divided by distance

        Velocity is equal to coefficient of permeability X hydraulic grad.

        Darcy's Law - Discharge equals cross-section area X Velocity

        Velocities range from 1/2 m/day to several meters/year

*Springs and Wells*

        Groundwater is obtained from springs or wells

*Springs*

        A flow of groundwater emerging naturally at the ground surface

        Change downward from an aquifer to an aquiclude causes spring

        Springs also occur in lava flows, joint systems, faults, etc.

*Wells*

        A hole in the ground that intersects the water table

        Cone of depression develops around a well as water flows in

        Wells occur in jointed Ig-Met rocks, but not unjointed areas

        Perched water body where an aquiclude traps water above table

# Aquifers

A highly permeably rock body in the zone of saturation

*Unconfined and confined Aquifers*

Unconfined - the upper surface is the water table

Confined - upper and lower surfaces are aquicludes

***Talk about the High Plains Aquifer and the Dakota Aquifer**

*Artesian Systems*

A sloping aquifer between 2 aquicludes, bottom closed off

*The Floridan Aquifer*

A complex system of confined and unconfined aquifers in ls.

133 km long, 19,000 years for water to traverse the distance

# Mining Groundwater

When groundwater is removed faster than it recharges

*Lowering of the Water Table*

When withdrawal exceeds recharge

Artificial recharge may abate lowering water tables

*Land Subsidence*

When water is withdrawn, mineral grains are unsupported

# Water Quality and Groundwater Contamination

Much groundwater is contaminated, naturally and artificially

*Chemistry of Groundwater*

Hard water - significant quantities of $Ca^{2+}$, $Mg^{2+}$, and $HCO_3^{1-}$

Soft water - few dissolved ions

May have variety of dissolved minerals, $H_2S$, sulfates, chloride

*Pollution by Sewage*

Sewage is most common source of groundwater pollution

Sand will readily filter biological pollutants

*Contamination by Seawater*

Excessive fresh withdrawal along sea coasts allows salt incursion

*Toxic Wastes and Agricultural Poisons*

Toxic leachates from domestic and industrial land fills

Major problem in most developed nations

Agricultural pesticides and herbicides are equally disastrous

*Underground Storage of Hazardous Wastes*

 Innumerable toxic and radioactive materials to be disposed of

 Underground disposal appears to be most stable site

 Groundwater circulation is greatest concern of this type disposal

 *** Go over the characteristics of an ideal disposal site***

 This calls for geologists to predict the future

## Geologic Activity of Groundwater

 Regions underlain by rocks subject to dissolution

*Dissolution*

 The process of minerals going directly into solution, carbonates

 Carbonates are highly soluble in weak carbonic acid

 $H_2O + CO_2 + CaCO_3 <-> Ca^{2+} + 2(HCO_3)^{1-}$ much excess water

 Weathering of carbonates leaves very little residue

 Carbonate weathering is much faster than other gradation types

*Chemical Cementation and Replacement*

 Minerals dissolved in groundwater precipitates as cement

 Replacement is volume to volume exchange of compounds

*Carbonate Caves and Caverns*

 Mans interest in caves dates into antiquity

 Caves come in all sizes from tiny to huge

 Cave formation is a chemical dissolution process

 Rate of cave formation increases with quantity is acid passing

 A few caves may have been dissolved by sulfuric acid

*Cave Deposits*

 Caves accumulate dripstone and flowstone deposits later

 Dripstone forms as water drips from cave surfaces

 Flowstone forms as water flows over cave surfaces

 Cave deposits form in partially air filled chambers

 Cave deposits may give evidence of past climates

*Sinkholes*

 A large dissolution chamber open to the sky

 Form when the cave roof weakens and collapses

 May form almost instantaneously by collapse

*Karst Topography*

An area underlain by carbonates, with much solution/collapse

Karst may also be underlain by dolomite, gypsum, salt

Most common in temperate to tropical, underlain by soluble rx

Sinkhole karst is most common landscape

Cone karst & tower karst in thick, well jointed limestone

Pavement karst most common where glaciated

## Toxic Groundwater in the San Joaquin Valley

Selenium is a necessary trace element in our diet

1983 selenium conc. in San Juaquin Valley became toxic

Irrigation water had leached selenium from ag. soils

Original source was marine sediments in Coast Range

Groundwater in western part of valley is alkaline, increasing sol.

Area is underlain by a clay aquiclude that complicates things

# OVERVIEW

Groundwater is one of the most valuable natural resources we have and at the same time one of the least understood. Most students will have some concept of groundwater because of the publicity it receives in the media. Sadly, they will hold many misconceptions, especially that water occurs in "lakes" and "channels" these are difficult to dispense and may require some effort.

Water table is a term many will be familiar with, but will need some explanation on its configuration and reasons why. A good discussion of types of well both ordinary and artesian will yield good results.

Groundwater as a renewable resource and as a mined quantity are extremely important to get across.

A topic that will fascinate the students will be groundwater erosion and land forms. This is an important spot for slides, also lots of fun.

# LEARNING OBJECTIVES

1. Know that the ultimate source of groundwater is surface infiltration.

2. Know why the zone of capillarity is present.

3. Be able to explain why the groundwater flow lines are down and out from a hill and up and in under a stream.

4. be able to list the variables that determine how long it takes for water to move from a recharge area to a discharge area.

5. Be able to explain why it is possible to determine discharge, velocity, or coefficient of permeability of groundwater passing through an aquifer if any two of the factors are known.

6. Know what the hydraulic gradient is and what its importance is in determining the rate of flow of groundwater.

7. Know why sandstones are generally better aquifers than siltstones or shales.

8. Know which features in igneous and metamorphic rocks promote the flow of groundwater through them.

9. Know how springs are related to the water table.

10. Know what causes a cone of depression to form around a producing well.

11. Know why the water level rises in an artesian well.

12. Know why sand is particularly effective in purifying water as it flows through the sand.

13. Know the meaning of the term "hard" water.

14. Know the origin of "hard" water in regions of carbonate rock.

15. Know why dripstone and flowstone form and how it is related to the zones of aeration and saturation.

16. Explain why karst landscapes are not well developed at high latitudes.

17. Explain the factors you would evaluate related to groundwater if you were picking a site for a garbage dump/landfill, cause you to pick a site or eliminate one.

## LECTURE DEMONSTRATIONS

1. The water table and zones of aeration and saturation can be demonstrated by filling a glass container with sand and then partially filling it with water. The addition of food coloring may help. Coarser sand will eliminate the capillary fringe.

2. Permeability and groundwater mining can be demonstrated with 2 or 3 clear (plastic) tubes that have been filled with different sizes of sand or silt. Water is poured onto the tops of the tubes to see how fast water percolates through each. You can easily demonstrate that water can be removed much faster that it can percolate through the column.

## FILMS AND VIDEO CASSETTES

1. Geysers and Hot Springs of Yellowstone, (source: IV).

2. Groundwater: The Hidden Reservoir, 19 min., 1971, (source: JW, IU).

3. Let's Explore a Cave, (source: IV).

4. Little Plover River Project: A Study in Sand Plains Hydrology, 32 min., 1963, (source: USGS).

5. Geology of Yellowstone, 16 min., 1967, (source: MGH).

6. Groundwater, 26 min., 1960, (source: UT).

7. Groundwater, 18 min., 1982, (source: EB).

8. Ground Water - A Part of the Hydrologic Cycle, 29 min., (source: ES).

9. Ground Water - The Hidden Reservoir, 19 min., 1971, (source: MG).

10. Ground Water: America's Buried Treasure, 14 min., (source: NWWA).

11. Ground Water: America's Hidden Reservoir, 18 min., (source: NWWA).

12. The Subject is Water, 28 min., 1976, (source: MTP).

13. The Water Below, 30 min., 1964, (source: MTP).

14. A Visit to the Petrified Forest, (source: IV).

## REFERENCES

Davis, S. N., and R. J. M. DeWiest, 1966, Hydrogeology, New York: Wiley.

Dunne, T., and L. B. Leopold, 1978, Water in Environmental Planning, San Francisco: Freeman.

Fetter, C. W., Jr., 1980, Applied Hydrogeology, Columbus, OH: Chas. E. Merrill Publ. Co.

Gardner, M., 1957, Fads and Fallacies in the Name of Science, New York: Dover.

Heath, R. C., 1983, <u>Basic Groundwater Hydrology,</u> U. S. geological Survey Water Supply Paper 2220.

Jackson, D. D., <u>underground Worlds,</u> Alexandria, VA: Time-Life Books.

Kazman, R. G., 1972, <u>Modern Hydrology, 2nd ed.,</u> New York: Harper and Row.

Meinzer, O. E., (ed.), 1942, <u>Hydrology,</u> New York: Dover.

Palmer, A. N., 1984, Geomorphic Interpretation of Karst Features *in* LaFleur, R. G. (ed.), <u>Groundwater as a Geomorphic Agent,</u> Boston: Allen and Unwin.

Price, M., 1985, <u>Introducing Groundwater,</u> Boston: Allen and Unwin.

Pye, V. I., and R. Patrick, 1983, Groundwater Contamination in the United States, <u>Science,</u> v. 221, pp. 713-718.

Sayer, A. N., Groundwater, <u>Scientific American,</u> v. 183, pp. 14-19.

Sweeting, M. M., 1973, <u>Karst Landforms,</u>New York: Columbia Univ. Press.

Todd, D. K., 1980, <u>Groundwater Hydrology, 2nd ed.,</u> New York: Wiley.

Trudgill, S., 1985, <u>Limestone Geomorphology,</u> New York: Longman.

Watham, T., 1975, <u>Caves,</u> New York: Crown.

# CHAPTER 11

## GLACIERS AND GLACIATION

### OUTLINE

**Vignette on "fossils" in glaciers**

    Fossils in and under glaciers give a feel for ice age events

    "Fossil" atmosphere suggests changes in ice age climates

**Glaciers**

    A permanent body of recrystallized snow moving by gravity

    Occur at high latitudes and altitudes, permanently cold

*Mountain Glaciers and Ice Caps*

    Cirque glacier - in a bowl-shaped mountainside depression

    Valley glacier - fills high valleys, may start as a cirque glacier

    Fjord glacier - occupies a deeply eroded coastal valley

    Piedmont glacier - a broad, bowl-shaped glaciers in front of mts.

    Ice cap - cover a mountainous highland, radial flow

*Ice Sheets and Ice Shelves*

    Continental sized ice masses, Greenland and Antarctica

    May cover entire land mass plus rest on the seafloor

    Ice shelf - a mass of floating ice

*Temperate and Polar Glaciers*

    Internal temperature is important in classification of glaciers

    Temperate glacier -at pressure melting point, ice and meltwater

    Polar glacier - below pressure melting point, solid ice

    Glaciers are warmed internally as melt water refreezes

*Glaciers and the Snowline*

    Glaciers form at,or above, the snowline

    Snowline is the lower limit of permanent snow, climate control

    Glaciers form wherever the snowline intersects the land surface

*Conversion of Snow to Glacier Ice*
> Glaciers are metamorphic rock with ice as the mineral
> Snow flakes sublime at the points and grow at the centers
> With deeper burial, porosity decreases and ice becomes solid
> Final density is 0.9 gm/cm$^3$
> Deeper burial causes the size of ice crystals to increase, 1+ cm

*Why Glaciers Change in Size*
> In general, glaciers have been getting smaller for a century

*Mass Balance*
> The input snow plus output melt water change with climates
> Accumulation is total snow input
> Ablation is total meltwater output
> Mass balance is difference between accumulation and ablation
> Accumulation area is upper part, ablation area is lower part
> Equilibrium line is the boundary and fluctuates with climate

*Fluctuations of the Glacier Terminus*
> If the mass balance is positive the terminus advances
> If the mass balance is negative the terminus retreats
> Terminus is stationary if mass balance is fixed

*Response lags*
> Terminus activity lags up to a decade behind climatic changes

*Calving*
> Retreat of a glacier that ends in the ocean, forms icebergs

*How Glaciers Move*
> Ice moves carrying objects with it
> Move by internal flow and sliding of basal ice over underlying

*Internal Flow*
> Movement by deformation within individual crystals
> Crevasses form when brittle surface has movement at depth

*Basal Sliding*
> Meltwater at base of temperate glaciers allows basal sliding
> Polar glaciers freeze to base and move internally only

### Velocities and Directions of Flow

Maximum rate at top center, minimum along sides and base

Ice in accumulation area moves down, in ablation area moves up

Whether the terminus advances - retreats, the mass moves down

Rates of movement are a few centimeters to few meters per day

### Glacier Surges

An unusually rapid movement and change in size and shape

May be related to hydrostatic water pressure

# Glaciation

The modification of the land surface by action of glacier ice

Includes erosion, transportation, and deposition of sediment

### Glacial Erosion and Sculpture

Acts as a plow, file, and sled

### Small-Scale Features of Glacial Erosion

Glacial striations, scratches caused by rocks frozen in base

Glacial grooves caused by large rocks dragging base

Glaciers smooth and polish the base, gentle slope up stream

### Landforms of Glaciated Mountains

Glaciers form distinctive alpine scenery

### Cirques

Bowl-like form caused by frost-wedging, plucking, abrasion

Tarn, a small lake in a cirque

Probably begin by ice wedging and flushing by meltwater

Arête forms as 2 cirques erode headward

Horn forms when 3+ cirques erode headward on a high peak

### Glacial Valleys

U-shaped cross section, bottom below tributaries

### Fjords

Valley carved below sealevel on the coast

### Landforms Produced by Ice Caps and Ice Sheets

### Abrasional Features

Striations shows center of ice caps were at pressure melting pt.

If mountains are not completely buried, tops are very rough

### Streamlined Forms

Drumlins, streamlined hill, gentle upstream, depo. or erosional

*Transport of Sediment by Glaciers*

        Load is multisized; suspended, drug throughout entire volume

        Load is concentrated along base and sides

        Much of basal load is fine crushed and ground - rock flour

*Glacial Deposits*

        Extremely efficient agent of erosion and transportation

        Collectively called glacial drift or just drift

        Deposits are generally neither sorted nor stratified

*Ice-Laid Deposits*

*Till and Erratics*

        Till is nonsorted drift deposited directly from ice, random mix

        Tillite is a lithified till

        Erratic is a rock different from underlying lithology

*Glacialmarine Drift*

        Sediment deposited on seafloor from ice shelves or icebergs

*Moraines*

        Sediment may be dropped directly or reworked by meltwater

        Moraine -deposits with form unrelated to underlying topography

        Ground moraine - widespread drift, smooth, 10+ m thick

        End moraine - ridgelike accumulation along the glacial margin

        Terminal moraine - ridgelike accumulation at end of glacier

        Lateral moraine - ridgelike accumulation along side of glacier

*Stratified Drift*

        Coarse to fine, turbulent meltwater to quiet lakes

*Outwash*

        Stratified sediment deposited in front of glacier by meltwater

        Outwash plains - braided stream deposits in front of glacier

        Valley train - meltwater deposits along valley walls

        Outwash terraces - formed as underloaded streams excavate

*Deposits Associated with Stagnant Ice*

        Ice-contact stratified till - rapid ablation, deposits between ice

## The Glacial Ages

        1821, European geologists suggested widespread glaciation

        1837, Louis Agassiz, hypothesis of continental glaciation

        Dramatic evidence of rapid global climate change, system respon

*Ice-Age Glaciers*

   During the Cenozoic the global climate has gradually cooled

   Last few million years numerous cycles of cold-warm

   30,000 years ago last great glaciers formed, moved equatorward

   About 29% of present land area was under glaciers

*Drainage Diversions and Glacial Lakes*

   Glaciers dam drainages causing lakes and rerouting of streams

*Lowering of Sea Level*

   Massive continental glaciers remove large volumes of seawater

   Removal of seawater lowers sealevel and moves shorelines

*Deformation of the Crust*

   Great piles of ice cause crust to subside

   Melting of ice allows studies of rates of isostatic crustal rebound

*Earlier Glaciations*

   Traditionally there have been 4 glacial ages described

   Deep-sea studies there have been 20 in last 3 m.y.

*Seafloor Evidence*

   Deep-sea studies have most complete Ice Age record

   Evidence from biologic remains as well as oxygen isotopes

*Pre-Pleistocene Glaciations*

   Early and late Proterozoic, early and late Paleozoic episodes

*Little Ice Ages*

   Minor episode during 16th and 17th centuries

## What Causes Glacial Ages?

   Cause - atmosphere, solid earth, oceans, biosphere, x-terrestrial

*Glacial Eras and Shifting Continents*

   Several ice ages are recorded in geol. record, tens of millions yrs

   1. movement of continents

   2. large-scale uplift due to continental collisions

   3. creation of mountain ranges during plate collisions

   4. opening and closing ocean basins by moving landmasses

   Glaciers occur at high latitudes and altitudes on windward side

   Position, shape, topography of continents change with time

   Mesozoic landmasses low alt., latitude, Cenozoic poleward, hi alt.

*Ice Ages and the Astronomical Theory*

      Ice ages have been going on for 3 m.y.

      Croll-Milankovitch, minor variations in orbital dist. and axial tilt

      1. axis of rotation wabbles

      2. tilt of axis of rotation varies

      3. eccentricity of orbit changes

      All of these determine the amount of radiation Earth receives

*Atmospheric Composition*

      Orbital factors give timing for ice ages

      Atmospheric factors exaggerate the orbital factors

      Air bubbles trapped in glaciers show decrease $CO_2$ and methane

      Amount of dust was much higher during ice ages

      Snow and ice cover would increase Earth's albedo

*Solar Variations, Volcanic Activity, and Little Ice Ages*

      Minor ice ages must have short term causes

      1. fluctuations in solar output

      2. large explosive volcanos with dust and gasses

## Glacial Water for Arid Lands

      As populations grow need for water grows--Antarctic icebergs?

      Is it possible or feasible to tow icebergs to site of need?

      Technological problems are formidable.  What unseen results?

## OVERVIEW

Glaciation is one of the fun and exciting topics for most students.  They are straight foreword methods of erosion and landscape sculpture and fairly easy for students to grasp.  Glaciers seem to occur in exotic places as opposed to streams that are everywhere.  This is another topic that requires copious use of slides.  It also can use imagination.  I remember when I lived in Vermont and trying to picture in my mind a pile of ice 10,000 feet thick lying to the north, I am not sure I ever really did.  Glaciers have also played a significant role in the development of human history.

The mechanics of snow-ice metamorphism is always a good reinforcement to earlier rock lectures.  Also, mechanics of glacial movement is easy to understand for most students.  Budgets is also easy and lends itself to analogies easily.

This is also a good chapter to show worldwide effects of fairly localized situations. It as is useful to illustrate natures influence on man.

Causes of glaciation is one of the great areas for conjecture in geology. It weaves numerous facts into relatively unprovable hypotheses.

## LEARNING OBJECTIVES

1. Know that glaciers are permanent bodies of moving ice that consist largely of recrystallized snow.

2. Know and be able to describe the several types of glaciers as described by their geometry - cirque glacier, valley glacier, fjord glacier, piedmont glacier, ice caps, ice sheets, and ice shelves.

3. Know that ice in a temperate glacier is at the pressure melting point, and liquid water exists at the base of the glacier.

4. Know that in a polar glacier ice is below the pressure melting point, and is frozen to the rock on which it rests.

5. Know that glaciers can form only at or above the snowline, which is close to sea level in polar regions and rises to high altitudes in the tropics.

6. Know that the mass balance of a glacier is measured in terms of accumulation and ablation.

7. Know that the equilibrium line separates the accumulation area from the ablation area and marks the level on the glacier where net gain is balanced by net loss.

8. Know that temperate glaciers move as a result of internal flow and basal sliding.

9. Know that in polar glaciers, which are frozen to their bed, motion is much slower and involves only internal flow.

10. Know that surges involve extremely rapid flow, probably related to excess amounts of water at the base of a glacier.

11. Know that glaciers erode rock by plucking and abrasion.

12. Know that glaciers transport rock debris chiefly at the base and sides and includes material from fine rock flour to large boulders.

13. Know that mountain glaciers erode stream valleys into U-shaped glacial valleys with cirques at their heads.

14. Know that fjords are excavated far below sea level by glaciers in high-latitude coastal regions.

15. Know that glacial drift is sediment deposited by glaciers and glacial meltwater.

16. Know that till is deposited directly by glaciers.

17. Know that glacial-marine drift is deposited on the seafloor from floating glacial ice.

18 Know that stratified drift includes outwash deposited by meltwater streams and ice-contact stratified drift deposited upon or against stagnant ice.

19. Know that ground moraine is built up beneath a glacier, whereas end moraines (both terminal and lateral) form at a glacier margin.

20. Know that during glacial ages, huge ice sheets repeatedly covered northern North America and Eurasia, causing the crust beneath the ice to subside and world sea level to fall.

21. Know that glacial ages have alternated with interglacial ages in which temperatures approximated those of today.

22. Know that studies of marine cores indicate that more than 20 glacial-interglacial cycles occurred during the Pleistocene Epoch.

23. Know that glacial eras in Earth history are related to the favorable positioning of continents and ocean basins, brought about by movements of lithospheric plates.

24. Know that the timing of glacial-interglacial cycles appear to be closely controlled by changes in Earth's precession, tilt, and orbital eccentricity, three factors that affect the distribution of solar radiation received at the Earth's surface.

25. Know that changes in the atmospheric concentration of carbon dioxide, methane, and dust may help explain the magnitude of global temperature lowering during glacial ages.

26. Know that climatic variations on the scale of centuries and decades have been ascribed to fluctuations in energy output from the Sun or to injections of volcanic dust and gases into the atmosphere.

# LECTURE DEMONSTRATIONS

1. Color slides, films, videos are almost essential to delivering lectures on glaciation. Excellent slides may be purchased from the several sources listed elsewhere in the supplement.

2. Dynamic models of glacial flow can be produced by using a mixture of molding plaster and water. Fleisher and Sales (1972) describe a successful method that produces excellent analogs of crevasse and flow patterns, icefalls, shears, falls, folds. When hardened, the models may be sectioned in order to view internal structures.

3. Silly putty can be used to demonstrate a material that may be brittle when struck or torn quickly, but be plastic when allowed to move slowly.

4. Plastic deformation can be demonstrated when it is well below freezing, or if a very large freezer is available. Set a large(10-25 lb.) block of ice between the ends of 2 tables and hanging a piece of piano wire with a heavy weight on each end over the top of the ice. The wire will gradually move through the ice and the ice will heal behind.

# FILMS AND VIDEO CASSETTES

1. <u>The Athabaska Glacier: A Case Study,</u>24 min., 1981, (source: MG).

2. <u>Evidence for the Ice Age,</u>19 min., 1965, (source: EB, BU, IU, USC).

3. <u>Glacial Geology - Interior Plains Region,</u> 28 Min., 1974, (source: ES).

4. <u>Glaciation,</u> 11 min., 1965, (source: NFBC).

5. <u>Glacier Legacy,</u> 28 min., 1973, (source: PU).

6. <u>Glacier's Depositional Features,</u> (source: IV).

7. <u>Glaciers Erosional Features,</u> (source: IV).

8. <u>Glaciers on the Move,</u> 11 min., 1073, (source: EB).

9. <u>Glaciers Shape Our Earth,</u> (source: IV).

10. <u>The Great Ice Age,</u> (source: IV).

11. <u>Mountain Glaciers,</u> 19 min., 1966, (source: OSU).

12. <u>The Rise and Fall of the Great Lakes,</u> 15 min., 1968, (source: NFBC).

13. <u>Story of Two Creeks,</u> 27 min., 1969, (source: UW).

14. <u>Time-Lapse Study of Glacial Flow,</u> 16 min., (source: UWP).

15. <u>Valley Glaciers,</u> 20 min., 1982, (source: FHS).

# REFERENCES

Bailey, R. H., 1982, <u>Glacier,</u> Alexandria, VA: Time-Life Books.

Bowen, D. Q., 1978, <u>Quaternary geology,</u> New York: Pergamon.

Bradley, R. S., 1985, <u>Quaternary Paleoclimatology: Methods of Paleoclimatic Reconstruction,</u> Boston: Allen and Unwin.

Coates, D. R.(ed.), 1974, <u>Glacial Geomorphology,</u> Binghamton, NY: State Univ. of NY.

Crowell, J. C., and L. A. Frakes, 1970, Phanerozoic Glaciation and the Causes of Ice Ages, <u>American Journal of Science,</u>v. 268, pp. 193-224.

Denton, G. H., and T. J. Hughes, 1981, <u>The Last great Ice Sheets,</u> New York: Wiley.

Drewry, D., 1986, <u>Glacial Geologic Processes,</u> London: Edward Arnold.

Embleton, C., and C. A. M. King, 1968, <u>Glacial and Periglacial Geomorphology,</u> New York: St. Martin's Press.

Flint, R. F., 1971, <u>Glacial and Quaternary Geology,</u> New York: Wiley.

Frakes, L. A., 1979, <u>Climates Throughout Geologic Time,</u> New York: Elsevier.

Lowe, J. J., and M. J. C. Walker, 1984, <u>Reconstructing Quaternary Environments,</u>New York: Longman.

Matsch, C. L., <u>North America and the Great Ice Age,</u> New York: McGraw-Hill.

Mitchell, J. M., 1968, Causes of Climatic Change, American Meteorological Society, Meteorological Monograph 8.

Nilsson, T., 1983, <u>The Pleistocene: Geology and Life in the Quaternary Ice Age,</u> Boston: Reidel.

Patterson, W. S. B., 1981, <u>The Physics of Glaciers, 2nd ed., New York: Pergamon Press.</u>

<u>Price, R. J., 1973,</u> Glacial and Fluvioglacial Landforms, New York: Hafner Publ.

Turekian, K., 1971, <u>The Late Cenozoic Glacial Ages.</u> New Haven, CT: Yale Univ. Press.

Washburn, A. L., 1973, <u>Periglacial Processes and Environments,</u> London: Edward Arnold.

Wright, H. E., and E. G. Frey, 1965, <u>The Quaternary of the United States,</u> Princeton, NJ: Princeton Univ. Press.

# CHAPTER 12

## WIND ACTION AND DESERTS

### OUTLINE

***Vignette on the dust bowl of the 1930's***
>    Severe drought in area unsuited for agriculture caused disaster
>    Improved agricultural practices have helped mitigate problems

### Wind as a Geological Agent
>    Mars is a desert with dust storms lasting months
>    Wind is most important as an erosive agent in deserts

*Planetary Wind System*

*Circulation of the Atmosphere*
>    Atmosphere circulates in convection cells due to uneven heating

*Coriolis Effect*
>    Earth rotation causes free objects to veer right in N. hemisphere
>    Causes 3 wind cells in each hemisphere
>    Trade winds blow 30° to equator, NE in N. hemi, SE in S. hemi.
>    At 20° to 30° air masses in both hemi. descend causing deserts
>    Westerly winds blow 30° to 60°, SW in N. hemi., NW in S. hemi.
>    Polar easterlies blow poles to 60° in both hemispheres
>    Polar front, irregular boundary of westerlies and polar easterlies

*Climate*
>    Nonuniform heating, Coriolis effect, dist. of land-sea, topography
>    Average weather plus variability of weather for a number of years

*Movement of Sediment by Wind*
>    Hurricanes can produce major wind erosion/destruction
>    Air is approximately 1/800 density of water
>    Generally only able to suspend sand, rarely pebbles

*Wind-Blown Sand*

        Surface creep begins at 16 km/hr(4.5 m/hr)

        Saltation at higher velocity

*Saltation*

        Process most common in dune areas

        Amount moved goes up dramatically with increasing velocity

        Seldom saltate more than 1 meter high

*Sand Ripples*

        A series of regularly spaced ripples caused of saltation of fines

*Wind-Blown Dust*

        Silt/clay sizes travel long distances by suspension

        Dustiest places on Earth coincide with major deserts

*Mobilization and Transportation of Dust*

        Dead air space 1 mm thick makes erosion of dust difficult

        Dust erosion needs help from saltating/rolling sand grains

        Dust may be transported thousands of km

*Dust Storms*

        Major large-scale transportation events in arid/semi-arid regions

        Most common during droughts or man caused activities

*Deposition of Dust*

        Deposition if:

        1. velocity and turbulence decrease

        2. collision with rough or moist surfaces

        3. particles aggregate and act like larger particles

        4. washed out of air by rain

        Vegetation(forests especially) is an excellent trap

        Topographic obstructions

        Coarser material settles first

*Detrimental Effects of Wind-Blown Sediment*

        Annually millions of dollars and significant loss of life

        Severe damage to crops, orchards, livestock

        Engines suffer major damage from dust grinding them up

        Reduce visibility and aggravate accidents

        Airborne dust can cause damage and disaster to aircraft

        Dust inhalation causes major medical problems

*Wind Erosion*

       Deflation, picking up and removal of small clasts

       Abrasion, impact by wind-driven grains

*Deflation*

       Most common in deserts with little vegetation and loose clasts

       Significant damage to dry lands laid bare by agriculture

       May remove sediment at extreme rates

*Deflation Hollows and Basins*

       Generally less than 2 km X 1-2 m deep

       Occasionally up to 50 m to 100 m deep

       Erosion to water table eventually stops deflation

*Desert Pavement*

       A nearly continuous covering of gravel left behind by deflation

*Abrasion*

*Ventifacts*

       Bedrock and/or stones polished and shaped by wind abrasion

*Yardangs*

       A large elongate, streamlined, wind-eroded ridge in a desert

*Eolian Deposits*

       Much more extensive than present day deserts, long travel dist.

*Dunes*

       A hill or ridge of sand deposited by wind

       Probably begin with some minor obstacle and its dead air space

*Dune Form and Size*

       Asymmetrical, windward <12°, leeward 33° to 34°

       Move up windward side by saltation, avalanche down leeward

       Windward angle < leeward, determined by wind speed, size sand

       Asymmetry is a good direction tool

       Cross strata slope in direction wind is blowing

       Typical maximum height is 30 to 100 m, 500 m maximum

*Dune Types*

    Five types of dunes:

    1. barchan

    2. transverse

    3. longitudinal

    4. star

    5. parabolic

*Dune Migration*

    Sand transfer from the windward to leeward causes migration

*Sand Seas*

    Vast track of shifting sand in large desert

*Loess*

    Thick, uniform layer of wind deposited silt with fine sand & clay

    Highly productive soils

*Characteristics*

    Distinctive as a wind deposited material

    1. uniform blanket over hills and valleys

    2. contains fossils of land plants and pulmonary animals

    Stands in vertical walls because of molecular attraction

*Origin*

    Deserts and floodplains of glacial meltwater streams, alluvial fans

*Glacial Loess of North America and Europe*

    Mostly of glacial origin:

    1. shape and composition typically glacial

    2. thickest downwind from glacial meltwater streams

*Dust in Ocean Sediments and Glacial Ice*

    Important component of deep-sea sediments

    Plumelike bodies extend seaward from major desert areas

    Also occurs in layers in glaciers and ice sheets

*Volcanic Ash*

    Large quantities of easily identified ejecta are wind transported

## Deserts

    Area with annual precipitation <250 mm, evaporation >250 mm

*Origins of Deserts*

    25% of land area

    Five types of deserts:

    1. most extensive, descending air masses at 20° to 30°

    2. continental interiors

    3. rainshadow on leeward side of mountain range

    4. coastal with cold upwelling currents off shore

    5. polar deserts

*Desert Climates*

    High temperature, low precipitation, high evaporation rates

    Daytime heating causes strong winds

## Surface Processes and Landforms in Deserts

    Topography is distinct because gradation process rates differ

*Weathering and Mass-Wasting in Deserts*

    Regolith is coarser because of deflation, steeper, angular slopes

*Mechanical Weathering*

    Mechanical weathering more dominant that chemical

    Coarser grain size produces steeper slopes

    Breakage along vertical joints causes steep cliffs

*Desert Varnish*

    A thin, dark, shiny coating($MnO$) formed after long exposure

*Desert Streams and Fluvial Landforms*

    Most desert surface is covered by stream beds and alluvial fans

    Most desert streams evaporate before reaching the sea

*Flash Floods*

    A sudden, swift flood that can transport large quantities of sed.

    Flash floods erode and redeposit material very rapidly

*Fans and Bajadas*

    Alluvial fans are interbedded alluvium and debris-flow material

    Fans typically begin at a canyon mouth

    Bajada is a series of coalescing adjacent alluvial fans

*Desert Lakes and Playas*

    Playa, lake bed in the bottom of a desert basin, dry most of year

    Commonly grayish to white due to contained salts

*Pediments*

    A broad, relatively flat surface, eroded across bedrock, thin alluv.

    Longitudinally it is slightly concave upward, shallow channels

    Meets the mountain at its top abruptly, headward erosion

*Inselbergs*

    Steep-sided mountains, ridges, isolated hills on a flat surface

    Homogeneous, resistant to erosion rock, long live

## Desertification

    Invasion of deserts into former semiarid areas

    Caused by both man and nature

*Desertification in North America*

    Overgrazing, excess. groundwater withdrawal, poor ag. practices

*Countermeasures*

    Long term improvement of land use practices in general

## Tectonic Desertification

    ***Discuss the evidence for tectonic desertification of Asia***

## OVERVIEW

Even though deserts are a climate type on Earth, many students lack an understanding of what desert is really like. You will probably need to spend time developing the ideas of what a desert is really like. The text describes a desert in terms of total precipitation and evaporation rates.

Most students are unfamiliar with global atmospheric circulation and that will need to be developed for them. These ideas will be needed in order for the students to understand why deserts exist and why there are different types of deserts.

Differences between humid climate and arid climate weathering  may also need explanation. Both rates and dominance of chemical vs. mechanical plays a part. Also some of the special depositional features can be discussed. Slides and other video materials will help comprehension.

## LEARNING OBJECTIVES

1. Know that unequal heating of the Earth by solar radiation sets up convective circulation in the atmosphere.

2. Know what the Coriolis is and how it breaks the equator-to-poleward circulation in both hemispheres into cells dominated, respectively by the tradewinds, westerlies, and polar easterlies.

3. Know that climate is the average weather of a place over a period of years, together with a degree of variability of that weather.

4. Know that the Earth's climates are influenced by the distribution of land and oceans, and by the surface topography.

5. Wind moves saltating sand grains close to the ground and suspended dust particles at higher levels and sorts them during the process.

6. Know that through deflation and abrasion, winds create deflation basins, desert pavements, ventifacts, and yardangs.

7. Know that dines originate where obstacles distort flow of air.

8. Know that dunes have steep slip faces and gentler windward slopes.

9. Know that dunes migrate in the direction of wind flow, forming cross strata that dip downwind.

10. Know that loess is deposited chiefly downwind from deserts and from the floodplains of glacier meltwater streams.

11. Know that loess is stable after deposition and is little effected by wind.

12. Know that airborne tephra deposited during explosive volcanic eruptions decreases in both thickness and grain size away from the source vent.

13. Know that hot deserts consitute about one quarter of the world's nonpolar land area and are regions of slight rainfall, high temperatures, excessive evaporation, relatively strong winds, and sparse vegetation.

14. Know that polar deserts occur at high latitudes where descending cold, dry air creates arid conditions.

15. Know that mechanical weathering, flash floods, and winds are especially effective geologic agents in deserts.

16. Know that fans, bajadas, and pediments are conspicuous features of many deserts.

17. Know that pediments are probably shaped by running water and are eroded surfaces across which sediment is transported.

18. Know that inselbergs form in relatively homogeneous, resistant rocks and may remain as persistent land-forms for millions of years.

19. Know that recurring natural droughts can lower the water table, cause high rates of soil erosion, and destroy vegetation, thereby leading to the invasion of deserts into nondesert regions.

20. Know that overgrazing, excessive withdrawal of groundwater, and other human activities can promote desertification.

21. Know that desertification can be halted or reversed by measures that restore the natural balance.

## LECTURE DEMONSTRATIONS

1. Samples of loess, sand, desert varnish, ventifacts, and other desert features are excellent prop. for lecture.

2. Slides and other videos will help make lectures more exciting and understandable for most students.

## FILMS AND VIDEO CASSETTES

1. Desert Landforms, 19 min., 1972, (source: AIS).

2. The Dust Bowl, 23 min., 1960, (source: MGH).

3. The Great Mojave Desert, 52 min., 1971, (source: FI).

4. Monuments to Erosion, 11 min., 1974, (source: EB).

5. Sand - The Desert in Motion, 11min.

6. Sand Dunes, 27 min., 1960, (source: UT).

7. The Work of the Wind, (source: IV).

8. The Work of Wind and Running Water, 19 min, 1969, (source: MG).

## REFERENCES

Ahbrandt, T. S., and S. G. Fryberger, 1982, Introduction to Eolian Deposits in Scholle, P. A., and D. Spearing, (eds.), Sandstone depositional Environments, Tulsa, OK: American Association of Petroleum Geologists.

Bagnold, R. A., 1941, The Physics of Blown Sand and Desert Dunes, New York:Methuen.

Brookfield, M. E., and Ahlbrandt, T. S., (eds.), 1983, Eolian Sediments and Processes, New York: Elsevier.

Glennie, K. W., 1970, <u>Desert Sedimentary Environments,</u> New York: Elsevier.

Greeley,R., and J. Iverson, 1985, <u>Wind as a Geological Process on Earth, Mars, Venus, and Titan,</u> New York: Cambridge Univ. Press.

Hadley, R. F., 1967, Pediments and Pediment-Forming Processes, <u>Journal of geological Education,</u>V. 15, pp. 83-89.

McGinnies, W. G., *et al.*, 1968,<u>Deserts of the World,</u>Tucson, AZ: Univ. Arizona Press.

McKee, E. D., 1966, Structures of Dunes at White Sands National Monument, New Mexico, <u>Sedimentology,</u> v. 7, pp. 1-69.

McKee, E. D., ed., 1979, A Study of Global Sand Seas, U. S. Geological Survey Professional Paper 1052.

Nairn, A. E. M., 1964, <u>Problems in Paleoclimatology,</u> New York: Wiley - Interscience.

Nilsen, T. H., 1982, Alluvial Fan Deposits *in* Scholle, P. A., and D. Spearing, (eds.), <u>Sandstone depositional Environments,</u> Tulsa, OK: American Association of Petroleum Geologists.

Nilsen, T. H. (ed.), 1985, <u>Modern and Ancient Alluvial Fan Deposits,</u> New York: Van Nostrand Reinhold Co.

Nickling, W. G.(ed.), 1986, <u>Aeolian Geomorphology,</u> Boston: Allen and Unwin.

Pye, K., 1987, <u>Aeolian Dust and Dust Deposits,</u>New York: Academic Press.

Rachocki, A., 1981, <u>Alluvial Fans,</u> New York: Wiley.

United Nations Secretariat, Conference on Desertification, 1977, <u>Desertification: Its Causes and Consequences,</u> New York: Pergamon Press.

Walls, J., 1980, <u>Land, Man, and Sand,</u> New York: MacMillan.

Wells, S. G., and D. R. Haragan, 1983, <u>Origin and Evolution of Deserts,</u> Albuquerque, NM: Univ. of New Mexico Press.

# CHAPTER 13

## THE OCEAN MARGINS

### OUTLINE

### *Vignette on coastline erosion and instability*

Homes built on a solid rock coastline is a great investment
>80% of U.S. shoreline show moderate to severe erosion
1990, 50% of U.S. population within 75 km of coast
By 2010, 75% of U.S. population within 75 km of coast
Coastal erosion plus potential sealevel rise could be disastrous

### Ocean Currents and Tides

Shorelines are in constant flux

*Surface Ocean Currents*

Broad, slow drifts of surface water, >50 to 100 m deep
Air flow over the water causes drag that moves the water
Westward movement under Trade Winds, east under Westerlies
Warm currents parallel to east coasts, cold parallel to west coast

*Tides*

Rhythmic rise and fall of ocean water due to gravitational bulges
Two bulges on opposite sides of Earth
Earth rotates under 2 bulges per day, 2 high, 2 low tides per day
Sun has only 1/2 the influence of the moon due to its distance
Most apparent along coasts, hardly noticeable in open sea
May produce rapid currents and depositional features

### Ocean Waves

Receive energy from winds blowing across them
Size depends on wind speed, distance, and duration

*Wave Motion*

 Fig. 13.4 shows dimensions of an unobstructed wave

 Each water molecule travels in a loop as wave form passes

 Diameter is equal to wave height

 Loop motion negligible at depth of one half wavelength

*Wave Base*

 One half wave length is effective wave base, max. about 300 m

 As depth decreases, orbits become flatter until back and forth

*Breaking Waves*

 Near shore wave drags, length shortens - height increase

 Front of wave steepens until it falls over - breaks

 Breaking form varies with the coast

*Surf*

 Wave activity between line of breakers and the shore

 Water piles up on shore and returns in sheets and rip currents

 Geologic work is mostly done by the direct action of surf

*Wave Refraction*

 Gradual change in direction of wave form due to bottom drag

 Converge on headlands, diverge and diffuse energy in bays

 Tendency over time is to form smoother, less indented coasts

## Coastal Erosion and Sediment Transport

 Zone of constant erosion and deposition

*Erosion by Waves*

 Surf zone plus above and below sea level

*Erosion Below Sea Level*

 Erosion is rare below 7 m, about 1.5X wave height

*Abrasion in the Surf Zone*

 Surf rubs and grinds shoreline wearing it and the clasts away

*Erosion Above Sea Level*

 Air trapped in cracks is compressed by wave impact

 Compressed air in cracks dislodges blocks

 Generally confined to 10 m +/- mean sea level

*Sediment Transport by Waves and Currents*

 Transported along shore or moved onto continental shelves

*Longshore Currents*

Oblique striking waves produce a longshore current component

Longshore currents move sediment along shore, change season

*Beach Drift*

Oblique swash moves sediment up beach then straight down

*Beach Placers*

Concentration of heavies by surf and longshore currents

Large variety of minerals, both modern and ancient beaches

*Offshore Transport and Sorting*

Sediment is repeatedly picked up and moved seaward, sorted

Sediments are effectively size sorted by waves and currents

## A Variety of Coasts

Combined effects of wave action on a type of shore material

*Beaches*

Wave-washed sediment along a coast in the surf zone, in motion

May be derived from nearby cliffs, or transported along shore

Fig. 13.12, foreshore, berm, backshore(dunes)

*Rocky (Cliffed) Coasts*

Fig. 13.13, wave-cut cliff and beach and depositional beach

Wave-cut cliff caused by surf cutting at base, then collapse

Wave-cut beach, the solid portion below eroded cliff, thin cover

Fig. 13.16, sea caves, arches, stacks - differential erosion

*Factors Affecting the Shore Profile*

Geometry/profile are compromise of constructive/destructive

Gravel beaches are steeper than sand beaches

More water is available to move sediment up than down beaches

Storms erode and narrow beaches, calm deposits and widens

*Major Coastal Deposits and Landforms*

Depositional forms distinctive, result of longshore currents

*Marine Deltas*

Deltas prograde when sediment is supplied faster than eroded

Deltas are long lived, complex features

*Spits and Related Features*

        Elongate sand or gravel ridge built into, ending in, open water

        Tombolo, sand or gravel ridge built between land and island

        Spits are controlled by seasonal current directions

        Bay barrier is a sand or gravel body built across a bay mouth

*Beach Ridges*

        Old berms built parallel to beach during storm episodes

*Barrier Islands*

        A long, narrow sandy island offshore and parallel to coast

        Lagoon is a water body behind a barrier island

        Probably built as sea level rose after last ice age

        One or more ridges of dune sand with successive shorelines

*Organic Reefs and Atolls*

        Built by small organisms of $CaCO_3$ in clear, 18+°C sea water

        Fringing reef is built attached, or very close to, land

        Barrier reef is built further from shore w/ lagoon between

        Atoll is a circular reef often on top of a submerged volcano

## Coastal Evolution

        Highly variable because numerous factor affect them

*Types of Coasts*

        Tend to be steep and rocky or flat and sandy

        Steep, rocky coasts are on the lead edge of lithospheric plate

        Flat, sandy coasts are on the edge with the trailing plate

        Imposed over the tectonic setting are, glaciation, drowning, etc.

        Rock type and local structure are also controls

*Effectiveness of Coastal Processes*

        Storms on west coast 45° to 60°, east coasts low latitudes

*Changing Sea Level*

        Long term changes due to:

            1. volume changes due to glacial waxing and waning

            2. lithospheric plate motions change basin volume

*Submergence*

        During last 3 m.y., numerous glacial sealevel changes, submerged

*Emergence*

Most higher than "normal" shore features from fewer glaciers

*Sea-level Cycles*

Many relict shores from of glacial lowering - interglacial rising

*Relative Movements of Land and Sea*

Unraveling of sea-level fluctuations difficult, both global and local

## Coastal Hazards

Constant hazard due to storms

*Storms*

Single storms may erode several meters of rock or sediment

85% of U.S. population live in 30 states with coastlines

Major storms seriously impact these people

*Tsunamis*

Dangerous seismic sea waves caused of seafloor disturbances

Great wave length/low amplitude in open sea

Short length/great amplitude as wave drags, slows near shore

*Landslides*

Result of surf erosion at toe of cliff

May cause giant sea waves that do immense damage

## Protection Against Shoreline Erosion

Oceanfront property is of great value

*Protection of Seacliffs*

Armor plate cliffs with boulders or build seawalls parallel to cliff

Very expensive, not effective to extraordinary storms

*Protection of Beaches*

Beaches are in delicate balance,any structures upset the balance

*Breakwaters and Groins*

Breakwaters are offshore protection to beaches and harbors

May alter straight beaches to scalloped beaches

Grions are walls perpendicular to a beach

These check beach drift and alter sand distribution

*Artificial Nourishment*

Sand hauled into upper end of a beach is a continuous expense

*Effects of Human Interference*
>Building dams upstream on rivers robs beaches of sand source

## How to Modify an Estuary
>**Discuss the changes on the San Francisco, or a local, estuary*
>Upstream water diverted for agriculture
>Increased urbanization destroys marsh lands
>Upstream hydraulic mining has caused great sediment influx
>Wildlife habitats and resources destroyed &/or polluted
>Sediment becomes toxic material sinks
>Most of world's large estuaries are being similarly impacted

# OVERVIEW

Shorelines are a topic that most students have experienced either first hand or via some media. Shorelines keep the attention of students with only a little effort on your part. Geologically, shorelines are an important topic of study because of the constant dynamics of the environment. This is also a environment of great biological interest.

Topics that easily capture the student's interest are wave formation, Coastal erosion and deposition, coastal landforms, beach equilibrium, coastal management, sea level changes, coastline development, man's interference and the results. These topics can use some video assistance, fortunately the is a variety available. Many case histories are available and many are well documented visually.

Sea level fluctuations can be used to show interplay of geological processes and man, It also related back to glacial geology and forward to plate tectonics.

# LEARNING OBJECTIVES

1. Know that huge wind-driven currents circulate clockwise in the Northern Hemisphere and counterclockwise in the Southern Hemisphere carrying warm water poleward and cold water equatorward.

2. Know that twice-daily tides are generated as the surface of the rotating Earth passes through tidal bulges on opposite sides of the planet.

3. Know that tidal bulges are produced by the gravitational attraction of the Moon and the Sun.

4. Know that fast-moving tidal currents cause movement of sediment in bays, straits, estuaries, and other restricted places along coasts.

5. Know that in deep water, waves have little or no effect on the seafloor.

6. Know that at a depth equal to half its wavelength, a wave can begin to stir up the bottom sediments.

7. Know that most of the geological work of waves is preformed by surf at depths of 10 m or less.

8. Know that wave refraction tends to concentrate wave erosion on headlands and to diminish it along the inner shores of bays.

9. Know longshore currents and beach drift can transport great quantities of sand along coasts.

10. Know that on gentle, sandy coasts the beach typically consists of a foreshore, berm, and backshore.

11. Know that on rocky coasts the shore profile includes a wave-cut cliff, a wave-cut beach, and a depositional beach.

12. Know that organic reefs form in tropical seas where water temperatures average at least 18°C.

13. Know that a fringing reef on a volcanic island may be transformed into a barrier reef and ultimately an atoll as the volcano subsides.

14. Know that the shape of coasts partly reflects the amount of energy available to erode and deposit sediment.

15. Know that rock structure and degree of erodibility help dictate the form of rocky coasts.

16. Know that nearly all coasts have experienced recent submergence due to postglacial rise of sea level.

17. Know that some coasts have experienced more complicated histories of emergence and submergence due to tectonic and isostatic movements on which are superimposed the worldwide sea-level rise.

18. Know that infrequent powerful storms, tsumani, and large landslides can pose significant threats to people and structures in the coastal areas.

19. Know that erosional damage to a shore cliff can be minimized by a seawall or an armor of boulders.

20. Know that beach erosion is a serious problem along many coasts, but beaches can be temporarily protected by a series of groins or by importation of sand.

# LECTURE DEMONSTRATIONS

1. Coastal landforms and many coastal processes can be demonstrated by slides and other types of video materials that may be purchased or rented from sources listed elsewhere in this supplement.

2. If a wave tank is available all sorts of demonstrations can be worked out. If time is available just turn the students loose to experiment.

# FILMS AND VIDEO CASSETTES:

1. Barrier Beach, 20 min., 1970, (source: AIMS).

2. The Beach, A River of Sand, 20 min., 1965, (source: EB).

3. The Earth - Coastlines, 11 min., 1969, (source: CFV).

4. Marine Erosion Processes: Cliffed Coasts, 11 min., 1969, (source: UC).

5. Portrait of a Coast, 29 min., 1980, (source: CO).

6. Shoreline Sediments, 39 min., 1972, (source: PSU).

7. Tsunami, 28 min., 1965, (source: NOAA).

8. Waves on Water, 16 min., (source: EB).

# REFERENCES

Bascom, W., 1964, Waves and Beaches, New York: Doubleday, Anchor Books.

Bird, C. F., 1969, Coasts, Cambridge, MA: M.I.T. Press.

Bird, E. C. F., 1984, Coast: An Introduction to Coastal Geomorphology, 3rd ed., New York: Oxford Univ. Press.

Bird, E. C. F., 1985, The World's Coastline, New York: Van Nostrand, Reinhold.

Black, J. A., 1986, Oceans and Coasts: An Introduction to Oceanography, Dubuque, IA: Wm. C. Brown.

Carter, R. W. G. (ed.), 1988, Coastal Environments, New York: Wiley.

Coates, D. R. (ed.), Coastal geomorphology, Binghamton, NY: State University of NY.

Davies, J. L., 1980, Geographical Variation in Coastal Development, 2nd ed., New York: Longman.

Dolan, R., B. Hayden, and H. Lins, 1980, Barrier Islands, American Scientist, v. 68, pp.16-25.

Fox, W. T., 1983, At the Water's Edge, Englewood Cliffs, NJ: Prentice-Hall.

Groen, P., 1967, The Waters of the Sea, New York: Van Nostrand, Reinhold.

Inman, D. L., and B. M. Brush, 1973, The Coastal Challenge, Science, v. 181, pp. 20-32.

King, C. A. M., 1972, Beaches and Coasts, 2nd ed., London: Edward Arnold.

Komar, P. D., 1976, Beach Processes and Sedimentation, Englewood Cliffs, NJ: Prentice-Hall.

Sacket, R., 1983, Edge of the Sea, Alexandria, VA: Time-Life Books.

Shepard, F. P., 1973, Submarine geology, 3rd ed., New York: McGraw-Hill.

Shepard, F. P., and H. R. Wanless, 1971, Our Changing Coastlines, New York: McGraw-Hill.

Sillen, L. G., 1967, The Ocean as a Chemical System, Science, v.156, pp. 1189-1197.

Smead, R. E., 1982, Coastal Landforms and Surface Features, Stroudsburg, PA: Hutchinson, Ross.

Stowe, K., 1983, Ocean Science, New York: Wiley.

Trenhale, A. S., 1987, The Geomorphology of Rock Coasts, New York: Oxford Univ. Press.

# CHAPTER 14

## DEFORMATION OF ROCK

### OUTLINE

***Vignette causes of rock deformation***

      Greece is being stretched, made longer, and twisted

      Greece 1 meter longer during last century

      Crustal rocks are being deformed

      Convective forces in meso- & asthenosphere deform lithosphere

      Twisting and bending are deep processes, exposed by erosion

      Inference on mechanisms of deformation from lab studies

**How Rock is Deformed**

      Laboratory studies gives most of what is known about mechanics

*Stress and Strain*

      Stress is used in rock studies rather than pressure

      Uniform stress - stress same in all directions, *i.e.* in a liquid/gas

      Confining stress - center of a rock body, same all around

      Differential stress - stress not equal in all directions

      Fig. 14.1, for 3 types of stress

      1. tensional stress - stretches rock

      2. compressional stress - squeezes rock

      3. shear stress - slippage and translation

      Strain refers the deformation of rock

*Stages of Deformation*

      Rocks subjected to increasing stress pass through 3 stages:

      1. elastic deformation, reversible, nonpermanent

      (elastic limit, point where change becomes permanent)

      2. ductile deformation, irreversible change in shape &/or volume

      3. fracture, irreversible, the object breaks

*Ductile Deformation Versus Fracture*

Brittle material deforms by fracture, little ductile deformation

Ductile material deforms by changing shape

Fig. 14.5A - brittle deformation; Fig. 14.5B - ductile deformation

Conditions controlling Brittle vs Ductile properties

1. temperature
2. confining stress
3. time and strain rate
4. composition

*Temperature*

Higher the temperature, the more ductile and less brittle

*Confining Stress*

Higher confining stress hinders formation of fractures, >ductile

*Time and Strain Rate*

If buildup is slow <u>and</u> prolonged brittle deformation is possible

Strain rate is the rate rock is forced to change shape/volume

Failure by fracture is common in upper-crustal rocks

Brittle-ductile transition is depth where ductile dominates

*Composition*

1. brittle minerals- quartz, garnet; ductile minerals- calcite, clay
2. presence of water make minerals more ductile

Ductile rocks - limestone, marble, shale, slate, phyllite, schist

Brittle rocks - sandstone, quartzite, granite, granodiorite, gneiss

## Deformation in Progress

Most deformation is too deep or too slow to observe

Observable types are:

1. abrupt fracture, few cm to m in minutes or hours
2. gradual ductile deformation, slow, steady motion

*Abrupt Movement*

Fault - fracture in rock along which movement has occurred

Large movements are the sum of many small movements

Movements are horizontal and/or vertical, max. about 15 m

Earthquakes are the result of sudden fault movements

*Gradual Movement*

      Deep faults have brittle deformation at surface, ductile at depth

      Most of the crust appears to be in at least vertical movement

## Evidence of Former Deformation

      Study of rock yields evidence of great vertical & horizontal move

      Structural geology is the area of geology that studies deformation

*Strike and Dip*

      First step in solving deformational history is tilt angle-direction

      Strike is the direction of a horizontal line on surface of a rock

      Dip is angle of slope from horizontal down to rock surface

      See Figs. 14.9, 14.10

*Geologic Maps*

      Fig. 14.11 is a geologic map, study strike-dip symbols

      Most of map area is covered by soil-vegetation

      Area between outcrops is interpreted in drawing a geologic map

*Deformation by Fracture*

      Near surface rocks are brittle and are fractured

      Joints are fracture where there has been not sliding

      Faults are fractures where there has been sliding

*Relative Displacement*

      Absolute movement is virtually impossible to determine

      Relative movement is all that can be determined at best

*Hanging Wall and Footwall*

      Hanging wall block is the block of rock above an inclined fault

      Footwall block is the block of rock below an inclined fault

*Classification of Faults*

      See Fig. 14.13 for classes of faults

      1. dip of the fault

      2. direction of relative movement

*Normal Faults*

  Tensional stresses that pull crust apart or push up from below
  Hanging wall block moves down relative to the footwall block
  Graben or rift- downdropped block between 2 parallel faults
  Half-graben - block hangs up on 1 side and drops on only 1 fault
  Horst - an up thrust block with faults on both sides
  Normal faults are very common, also grabens and horsts

*Reverse Faults and Thrust Faults*

  Both arise from compressional faults
  Crust is shortened in both cases
  Hanging wall block moves up relative to the footwall block
  Thrust faults are very low angle reverse faults
  Common to great mountainous areas

*Strike-slip Faults*

  Principal movement is horizontal, parallel to strike
  Right lateral movement, when opposite side moves to your right
  Left lateral movement, when opposite side moves to your left
  See Fig. 14.18 about transform faults, all strike-slip

*Evidence of Movement Along faults*

  Movement along a break in rock is often difficult to determine
  Slickensides  are striated, polished surfaces after movement
  Fault breccia - ground, crushed rock next to a fault

*Deformation by Bending*

  Fold - any sort of a bend in rocks
  Folds result from ductile deformation, below brittle-ductile trans

*Types of Folds*

  Monocline, local steepening on otherwise uniformly dipping rx
  Anticline, upfold in the form of an arch
  Syncline, downfold in the form of a trough

*Geometry of Folds*

  ***See Fig. 14.21 and describe the parts of a fold***
  Limbs
  Axis
  Plunging fold
  Plunge
  Axial plane

***See Fig 14.22 and describe types of folds***

Open fold

Isoclinal

Overturned

Recumbent

Sedimentary features are useful in structural interpretations

*Relationship Between Folds and Faults*

Faults often die out as folds, folds die out to undisturbed rocks

Same stress will produce different brittle/ductile deformations

Thrust faults may be recumbent folds that break and slide

*Folds and Topography*

Topography is influenced by resistance to erosion plus structure

## Rock Deformation and Oil Pools

Most large oil fields are controlled by geologic structures

Five requirements for on oil pool to form

1. Source rock

2. Permeable reservoir rock

3. Impermeable roof rock

4. Reservoir and roof rock must form a trap

5. Trap must form before oil/gas so the cannot escape

Great oil fields of Middle East are all structural traps

# OVERVIEW

Rock deformation is one of the fun chapters of the text, but alas one that requires abstract thinking on the part of the students. Slides and models are essential for lecturing this chapter. One of the problems with structural geology and the next several chapters is the scale of the features. An introduction to geologic maps of the local area may prove helpful if the students can see what is actually on the ground and compare it with the map.

The text goes into some depth with deformation of rocks and stress - strain. This will require considerable lecture effort on your part. You will have to decide how much your students can comprehend.

Students seem to really get into structures and the economic geology that is related to them. Use lots of slides and have fun!!

# LEARNING OBJECTIVES

1. Know that rocks can be deformed in three ways: elastic deformation(no permanent change), ductile deformation(folds), and fracture(faults, joints).

2. Know that high confining stress and high temperatures enhance ductile properties.

2. Know that low confining stress and low temperatures enhance elastic properties and failure by fracture when the elastic limit is exceeded.

3. Know that the rate at which a solid is deformed (strained) controls the style of deformation.

4. Know that high strain rates lead to fractures.

5. Know that low strain rates cause folding.

6. Know that weak rocks (limestone, marble, slate, phyllite, and schist) enhance ductile properties.

7. Know that strong rocks (sandstone, quartzite, and granite) enhance brittle properties.

8. Know that dry rocks are stronger that wet rocks.

9. Know that orientations of contacts between strata, faults, joints, or any other inclined planar surfaces are described by the strike( compass direction of the intersection of the inclined surface and a horizontal plane), and the dip ( the angle measured down between the inclined surface and the horizontal plane).

10. Know that fractures in rocks along which slippage occurs are called faults.

11. Know that normal faults are caused by tensional stresses that tend to pull the crust apart.

12. Know that reverse and thrust faults are caused by compressional stresses that squeeze, shorten, and thicken the crust.

13. Know that strike-slip faults are vertical fractures that have horizontal motion and are caused by shear stresses.

14. Know that it is usually only possible to determine relative motions of rocks on either side of a fault.

15. Know that ductile deformation of strata causes bends or warps.

16. Know that folds are due to compressional stress.

17. Know that an upward arched fold is an anticline and a downward, troughlike fold is a syncline.

18. Know that the sides of a fold are called limbs.

19. Know that overturned folds (meaning both limbs dip in the same direction) are common in mountain ranges formed by continental collision.

20. Know that recumbent folds are overturned folds that have nearly horizontal limbs.

21. Know that faults die out by becoming folds, and folds die out by becoming smaller and smaller wrinkles.

## LECTURE DEMONSTRATIONS

1. Wooden or plastic block models may be purchased for any of several suppliers or you and/or your students may make them. Ones that show outcrop patterns before and after erosion are useful. Models from styrofoam can be taken apart.

2. If a squeeze box is available you can demonstrate faulting and folding in class for your students.

3. Silly Putty can be used to demonstrate brittle and plastic deformation as it relates to rate of deformation.

4. Folds and faults constructed from different colors of modeling clay can be cut, "eroded", at various angles to give map patterns and show internal structure.

5. Large samples and slides can be used to show slickensides, fault breccia, mylonite, and all sorts of other structural features.

6. Geologic maps of your local area are useful for relating structures on maps and on the ground. Some places you will have to go on an imaginary trip with slides or videos you make yourself to show what the ground looks like that is on the geologic map.

## FILMS AND VIDEO CASSETTES

1. How Solid is Rock?, 21 min., 1967, (source: EB, BU, IU, USC).

2. Faults and Folds, part I - 27 min., part II - 29 min., 1960, (source: UT).

3. The Making of a Continent, 1983 and 1986, segments "Collision Course" and "The Land of the Sleeping Mountains", (source: FI).

4. The San Andreas Fault, 21 min., 1974, (source: EB).

5. Why Do We Still have Mountains?, 20 min., 1965, (source: EB).

## REFERENCES

Anderson, D., 1971, The San Andreas Fault, Scientific American, v. 225, pp. 52-68.

DeJong, K. A., and R. Scholten, 1973, Gravity and Tectonics, New York: Wiley.

Dennis, J. G., 1987, Structural Geology - An Introduction, Dubuque, IA: Wm. C. Brown.

Hobbs, B. E., W. D. Means, and P. F. Williams, 1976, An Outline of Structural Geology, New York: Wiley.

King, P. B., 1969, The Tectonics of North America - A Discussion to Accompany the Tectonic Map of North America, Scale 1:5,000,000, U. S. Geological Survey Professional Paper 628.

King, P. B., 1977, The Evolution of North America, 2nd ed., Princeton, NJ: Princeton Univ. Press.

Ramsey, J. G., 1967, Folding and Fracturing of Rocks, New York: McGraw-Hill.

Shelton, J., 1966, Geology Illustrated, San Francisco: Freeman.

Suppe, J., 1985, Principles of Structural Geology, Englewood Cliffs, NJ: Prentice-Hall.

# CHAPTER 15

## EARTHQUAKES AND THE EARTH'S INTERIOR

### OUTLINE

*Vignette on earthquakes*

      Collapsing buildings kill people, earthquakes do not

      Damage and loss of life is greatest in non-reinforced buildings

      Lithospheric structure and earthquakes are coupled

      Study of seismic waves yields understanding of Earth's interior

### Earthquakes

      Result of stored elastic energy suddenly being released

      Sudden slippage of a fault causes vibrations to move thru Earth

      Elastic rebound theory is most widely accepted for earthquake

*Origin of Earthquakes*

      Elastic rebound theory suggests energy may be stored  on faults

      Energy may be released a little at a time or in great quantities

      most earthquakes occur in the brittle portion of the crust

*How Earthquakes Are Studied*

      Seismology - the study of earthquakes

*Seismographs*

      Seismograph - device to record earthquake shocks/vibrations

      Many seismographs use the principal of inertia

      Made to record both vertical and horizontal displacement

      Benioff strain seismograph measures strain in a silica-glass tube

      Vibration as small as $10^{-8}$ cm can be measured

*Earthquake Focus and Epicenter*

      Focus in the actual spot where the energy was released

      Epicenter is on the Earth's surface directly above the focus

## Seismic Waves

Earthquakes produce seismic waves

Seismic waves must be recorded during the disturbance

Body wave move through the main body of the Earth

Surface waves move in the near surface portion of the Earth

## Body Waves

P (primary) waves, fastest, compressional waves, body waves

S (secondary) waves, intermediate, shear waves, body waves

Both P and S waves may change direction by:

1. Reflection - bounce off of the surface of a different medium

2. Refract - change direction going into a different medium

Seismic body wave velocity a function of density of medium

## Surface Waves

Move much slower than P and S waves

Longer wavelengths penetrate deeper than short wavelength

Dispersion - the property of travel at different depths

## Location of the Epicenter

Utilizes the difference in time of arrival of the P and S waves

Draw circles from 3 seismograph stations distance to epicenter

Depth is determined from difference in P - S travel times

## Magnitude of Earthquakes

Large earthquakes are rare, small one are common

Very large earthquakes occur about once per century at a site

## Richter Magnitude Scale

Defined by the max amplitude of P and S waves 100 km from epi

Each step in magnitude is 10 fold wave amplitude increase

Fig. 15.10 shows how magnitude is calculated

Energy release as approx. the square of the amplitude increase

Largest earthquake ever recorded is 8.6 Richter magnitude

## Earthquake Risk

Largest quake to occur in U.S. was New Madrid, Mo, 1811-12

Fig. 15.11, NOAA map of seismic risk map

Fig. 15.12, USGS map of seismic risk map w/gravity acceleration

*Earthquake Disasters*

> In earthquake prone areas buildings are constructed to survive
>
> Were buildings are not constructed to code, disaster strikes
>
> Table 15.1 - 17 earthquakes causing 50,000 + deaths

*Earthquake Damage*

> 1. ground motion damages or destroys structures
> 2. building set on faults are split by motion
> 3. fire as cooking/heating devices break and fall
> 4. mass-wasting in steep regions
> 5. liquifaction of waterlogged soils
> 6. tsunami, seismic sea waves

*Modified Mercalli Scale*

> Based upon sensation of motion felt and structure damage

## World Distribution of Earthquakes

> Several well defined seismic belts:
>
> 1. circum-Pacific
> 2. Mediterranean-Himalayan
> 3. midocean ridges
>
> Also midocean ridges, deep trenches, andesitic volcanos
>
> Most foci at plate edges >100 km, a few at trench up to 700 km

*Benioff Zones*

> Deep earthquake foci paths beneath trenches to 700 km depth

*First-Motion Studies of Earthquakes*

> Study of seismograms indicates direction of fault initial motion
>
> Several foci sites are needed to determine strike-dip of a fault
>
> Strike-dip will allow distinction of up-down vs back-forth motion
>
> Both P and S waves can be used

## Earthquakes and Plate Tectonic

> Most detailed proof of plate tectonic is quake site + first motions
>
> Three kinds of plate boundaries:
>
> 1. divergent boundaries or spreading centers
> 2. transform fault boundaries
> 3. convergent boundaries

*Spreading Centers*

    Plates move apart, tension, normal faults, low magnitude

    Shallow focus, low magnitude quakes common

*Transform Fault Boundaries*

    Hugh, vertical strike-slip faults, plates slide past each other

    Shallow focus, high magnitude quakes common

*Convergent Boundaries*

    Subduction of oceanic crust capped lithospheric plate

    These are thrust fault boundaries, complicated

    Shallow to deep foci, generally low magnitude quakes

    Collision of continental crust capped lithospheric plates

    A collision zone several hundred km wide

    Foci up to 300 km deep, high magnitudes common, thrusting

# Earthquake Prediction

    Cause of many dreadful disasters, prediction a primary focus

    Reports of strange animal behavior preceding quakes

    Much research on properties of elastically strained rocks

    1. rock magnetism

    2. electrical conductivity

    3. porosity

    4. changes in well water level

    5. tilting of ground

    6. slow rises and falls of land surface

    7. small cracks and fractures-foreshocks

    In quake prone areas, patterns may be discerned

    Seismic gaps, areas where no quakes have occurred, but should

# Using Seismic Waves as Earth Probes

    Show rock property changes, distinct rock boundaries

*Layers of Different Composition*

    Wave velocity controlled by density and rigidity of a solid

    Wave study suggests neither comp. nor phys. props. are constant

    Discontinuities are detected by refraction and reflection

*The Crust*

    Mohorovicic described the discontinuity at base of Crust, 1900's

*Thickness and Composition of the Crust*

Oceanic crust is <10 km thick, basalt and gabbro

Continental crust is 20 to 60 km, granite and diorite

*The Mantle*

Probably rich in minerals olivine and pyroxene

Kimberlite pipes studies substantiate this conclusion

*The Core*

At 2900 km a major discontinuity

P-waves strongly reflected and refracted, S-waves dampened

Outer Core must be liquid

Composition must be iron because of the density also meteorites

*The Inner Core*

P-wave reflections indicate a solid inner core of iron

Solid because of the immense pressure

*Layers of Differential Physical Properties in the Mantle*

Fig. 15.26 indicates velocity changes in mantle

Composition of mantle appears uniform, changes are physical

*The Low-Velocity Zone*

100 to 350 km depth P and S waves slow down boundaries fuzzy

Zone is less rigid, less elastic, more ductile

Temperature is near close to onset of partial melting

Low-velocity is the asthenosphere of plate tectonic theory

*The 400 km Seismic Discontinuity*

Sudden velocity increase at 400 km

Polymorphic transition of olivine to spinel, more dense packing

*The 670 km Seismic Discontinuity*

Major increase of P waves, mantle density 10%

Diffuse boundary, may be polymorphic transition of pyroxene

May be related to subducting lithospheric plates

Seismic tomography gives not definitive answer

## Gravity Anomalies and Isostacy

Earth is slightly flattened at poles and bulges at equator

The pull of gravity does not change uniformly as predicted

*Gravity Anomalies*

        Gravimeters can measure gravity variations of one part in ($10^{-8}$)

        After corrections great anomalies due to rock differences exist

        Mountain ranges have roots that stick down into mantle

        Great thicknesses of low density mountain rock give neg. anom.

        Topographic relief of crust is because mantle is weak & buoyant

*Isostacy*

        Property of floatational balance among lithospheric segments

        Ice from Ice Ages depressed the crust and it is still rebounding

        Continents stand high because they are thick, lower density

        Ocean basins lie low because they are thin, higher density

        Isostacy is the principal explanation for vertical surface motions

## A Great Earthquake in the Pacific Northwest?

        Many of greatest earthquakes ever, recorded in subduction zones

        Foci in subduction zones generally within lithosphere <7.5

        Great quakes, 8.0+ have foci at the boundary of the 2 plates

        No major quakes along subducting plate from N. Cal. to Cent. B.C.

        Is the region due for some huge quakes?

        Evidence still not conclusive either way

## OVERVIEW

Earthquakes are a topic that evokes both interest and terror to the average person. Every student has some ideas concerning the topic.. The ideas may be correct, or wildly incorrect. Modern news reporting has brought earthquakes into every home in America because even moderate sized earthquakes make good copy. Modern movie makers have also brought them to "life". Part of your job as lecturer of this course will be to provide a scientific basis for what is often incorrectly portrayed. Treat the students carefully, feelings can be hurt and science become taboo.

Earthquakes and their causes and results is of great interest. Students always enjoy hearing about great disasters and what really made them disasters, what happened after the earthquake.

Uses for earthquake data leads into their distribution and into plate tectonics. Study of the internal structure of the Earth is another natural out growth of this topic.

# LEARNING OBJECTIVES

1. Know that earthquakes are caused by abrupt movements along faults as elastic energy is released.

2. Know that earthquake vibrations are measured with seismographs.

3. Know that energy released at an earthquake focus radiates outward as body waves.

4. Know that there are two types of body waves - P waves (**P**rimary waves that are compressional) and S waves (**S**econdary waves that are shear waves).

5. Know that earthquake energy also causes the surface of the Earth to vibrate because of surface waves.

6. Know that the epicenter is the point on the Earth's surface directly above the focus.

7. Know that the focus and epicenter of an earthquake can be located by measuring the differences in travel times between P and S waves.

8. Know that the amount of energy released during an earthquake is calculated on the Richter magnitude scale using seismograph records of the seismic body waves.

9. Know that earthquakes are not uniformly distributed over the Earth: the circum-Pacific belt (80%), the Mediterranean-Himalayan belt (15%), remainder (midocean ridges, etc.) of the Earth (5%).

10. Know that seismic belts outline tectonic plates.

11. Know that direction of movement across a plate boundary is determined from seismic-wave first-motion studies.

12. Know that seismic body waves can be refracted and reflected just as sound and light waves are.

13. Know that the study of seismic-wave refraction and reflection has lead scientists to infer the internal structure of the Earth by locating boundaries, or discontinuities, in its composition and physical properties.

14. Know that pronounced compositional boundaries occur between the crust and the mantle and between the mantle and outer core.

15. Know that the base of the crust is a pronounced seismic discontinuity called the Mohorovicic discontinuity.

16. Know that the thickness of the crust ranges from 20 to 60 km in continental regions but is only about 10 km beneath the oceans.

17. Know that within the mantle there are two zones, at depths of 400 and 670 km, where sudden density changes produce seismic-wave discontinuities.

18. Know that the 400 km change is probably produced by a polymorphic transition of olivine.

19. Know that the 670 km change is due either to a polymorphic transition or/and a compositional change.

20. Know that the core has a high density and is inferred to consist of iron plus small amounts of nickel and other elements.

21. Know that the outer core must be molted because it does not transmit S waves.

22. Know that the inner core must be solid because of the way P waves react as they travel through.

23. Know that from a depth of 100 km to a depth of 350 km there is a zone of low seismic-wave velocity that also causes pronounced dispersion of surface waves.

24. Know that the low seismic-wave velocity zone coincides with the asthenosphere.

25. Know that the lithosphere, the rigid, 100 km thick outer layer of Earth, overlies the asthenosphere.

26. Know that the outer portions of the Earth are in approximate isostatic balance; in other words, like huge icebergs floating in water, the lithosphere "floats" on the asthenosphere.

## LECTURE DEMONSTRATIONS

1. Elastic deformation can be nicely demonstrated by sliding two fairly thick pieces of foam rubber past each other. If you draw different colored, but matching, lines on the two pieces, bending shows up nicely. Lessening deformation with distance shows up also. Vibrations (earthquakes) during the rupture (sliding) can be easily felt.

2. The difference between P and S waves can be demonstrated using a "SLINKY" with a rope threaded lengthwise.

3. Slides and video materials are available for several sources listed elsewhere in the supplement.

4. Copies of seismograms, seismic maps, and slides of earthquakes are available from NOAA, National Geophysical Data Center, E/GC1. 325 Broadway, Boulder, CO 80303.

## FILMS AND VIDEO CASSETTES

1. After the Earthquake, 24 min., 1982, (source: UM).

2. The Alaskan Earthquake, 22 min., 1966, (source: NAVC).

3. Alaskan Earthquake, 20 min., 1964, (source: MTP, IV).

4. An Approach to the Prediction of Earthquakes, 27 min., 1967, (source: AEF).

5. Earthquake!, 14 min., 1965, (source: BU, IU, USC).

6. Earthquake Prediction and Prevention, 24 min., 1976, (source: MG).

7. Earthquake, Lesson of a Disaster, 13 min., 1971, (source: EB).

8. Earthquakes: Exploring the Earth's Restless Crust, 22 min., 1983, (source: EBE).

9. A Predictable Disaster, 32 min., 1976, (source: TLV).

10. The Restless Earth, Part 3, 27 min., 1972, (source: IU).

11. The San Andreas Fault, 20 min., 1974, (source: EB, IU).

12. San Francisco: The City That Waits to Die, 57 min., 1971, (source: TLV).

13. The Trembling Earth, 30 min., 1968, (source: IU).

14. Tsunami, 28 min., 1965, (source: NOAA).

15. When the Earth Moves, 24 min., 1976, (source: USGS, MTP).

## REFERENCES

Bolt, B. A., W. L. Horn, and G. A. Macdonald, 1977, Geological Hazards 2nd ed.,New York: Springer-Verlag.

Bolt, B. A. (ed.), 1980, Earthquakes and Volcanos, readings from Scientific American, San Francisco: Freeman.

Bolt, B. A., 1988, Earthquakes, San Francisco: Freeman.

Bott, M. P. H., 1971, _The interior of the Earth,_ New York: St.Martin's Press.

Coulomb, J., and G. Jobert, 1973, _The Physical Constitution of the Earth,_ Edinburgh: Oliver and Boyd.

Elby, C. A., 1980, _Earthquakes,_ New York: Van Nostrand, Reinhold.

Hill, M., 1981, San Andreas Fault: History of Concepts, _Geological Society of America Bulletin,_ v. 92, pp. 112-131.

Isacks, B., J. Oliver, and L.R. Sykes, 1968, Seismology and the New Global Tectonics, _Journal of Geophysical Research,_ v. 73, pp. 5855-5899.

Okamoto, S., 1973, _Introduction to Earthquake Engineering,_ New York: Halsted.

Press, F., 1975, Earthquake Prediction, _Scientific American,_ v. 232, p. 14-23.

Raleigh, C. B., _et al.,_ 1976, An Experiment in Earthquake Control at Rangley, Colorado, _Science,_ v. 191, pp. 1230-1236.

Rankin, D. W., 1977, _Studies Related to the Charleston, South Carolina, Earthquake of 1886,_ U. S. Geological Survey Professional Paper 1028. (also see Professional Paper 1313).

Rikitaka, T., 1976, _Earthquake Prediction,_ New York: Elsevier.

Stacey, F. D., 1977, _Physics of the Earth, 2nd ed.,_ New York: Wiley.

U. S. Geological Survey, _The Alaskan Earthquake,_ Professional Papers 541 (1965) thru 546 (1970).

U. S. Geological Survey, _The San Fernando, California, Earthquake of February 9, 1971,_ Professional Paper 733.

Utgard, R. O., G. D. McKenzie, D. Foley, 1978, _Geology in the Urban Environment,_ Minneapolis, MN: Burgess Publishing Co.

Walker, B., 1982, _Earthquake,_ Alexandria, VA: Time-Life Books.

Wyss, M., 1979, _Earthquake Prediction and Seismic Patterns,_ Basel, Switzerland: Birkhauser Verlag.

# CHAPTER 16

## GLOBAL TECTONICS

### OUTLINE

***Vignette on ideas why the Earth's surface is configured as it is***

    Why are the continents and ocean basins shaped as they are?

    Why do earthquake belts, mountain ranges, volcanos occur?

    19th century - originally molten Earth cooling and contracting

    Early 20th century an expanding Earth

    Mid - Late 20th century, plate tectonics

### Continental Drift

    1912, Alfred Wegener, continents drift slowly across globe

    Collide to form supercontinents, tear apart to form small masses

*Pangaea*

    Attempt to explain why coastlines on 2 sides of Atlantic match

    Supercontinent, Pangaea, during Permian

    Mesozoic Pangaea broke into Laurasia and Gondwanaland

    Evidence, Permian glaciation in southern hemisphere

    Problem, no mechanism to slide the continents around

*Apparent Polar Wandering*

    Mid 1950's - mid 1960's many discoveries

    Earth's "fossil" magnetic record, 3 pieces of information

        1. Earth's polarity - whether normal or reversed

        2. location of poles at time of formation of the rocks

        3. magnetic latitude of rock at time of formation

    Apparent polar wandering - magnetics indicates poles moved

*Seafloor Spreading*

    If the continents move, shouldn't the oceans move also?

    1962, Harry Hess, suggested the hypothesis-Seafloor Spreading

    Suggested magma rose from mantle to form new ridge seafloor

Test proposed by Vine & Matthews and by Morley

Suggested test was magnetism of oceanic crust

Cooling lava assumes magnetism of global magnetic

Seafloor should preserve a history of magnetic reverses

Rate of spreading could also be determined - up to 10 cm/yr

*Plate Tectonics: A New Paradigm*

Seafloor spreading was spur to development of Plate Tectonics

1. study of low-velocity zone in mantle

2. rigid lithosphere was strong enough to be coherent

New seafloor is created, old floor must be destroyed-Benioff Zone

*Structure of a Plate*

Six large, numerous small plates all about 100 km thick, rigid

Margins are active zone, interiors are stable

Margin movements are: divergent, convergent, transform

*Divergent Margins*

Spreading center, newly formed, newly magnetized rock

*Magnetic Records and Plate Velocities*

Most recent reversal 730,000 yr ago, oldest 165 m.y., Jurassic

Relative velocities are equal on both side of the spreading center

*Variations in Plate Velocities*

Plates with much continental crust move slowest, no cont. fast

Rates of motion vary in relationship to the spreading pole/axis

1. width of new oceanic crust increases with distance from pole

2. projection of spreading center passes thru spreading pole

3. each transform fault is a line of spreading pole latitude

*Topography of the Seafloor*

Controlled by growth and movement of plates

1. mid-ocean ridges, shape controlled by rate of spread

2. seafloor adjacent to ridges are high, subside as they cool

*Convergent Margins*

Subduction zones, lithosphere subsides into interior

Interpretation based upon:

1. seismic studies

2. deformed rocks where 2 convergent plates meet

3. associated volcanism

*Magmatic Arcs*
>    Descending plates are heated, wet partial melting, andesitic mag
>    Magmatic arc, parallel to, and 100 to 400 km from, trench
>    May form a island arc, or continental island arc
>    Old, cold crust descends steeply, steeper Benioff Zone
>    Younger, "warm" crust descends less steeply, flatter Benioff Zone
>    Earthquakes as long as crust is brittle, up to 700 km deep

*Mélange*
>    A chaotic mixture of broken, jumbled, and thrust-faulted rock
>    May have high-pressure, low-temperature metamorphism

*Fore-Arc Ridges and Fore-Arc Basins*
>    Fore-Arc Ridge is underlain by mélange, locally thickened crust
>    F-Arc Basin, low-lying area between fore-arc ridge/magmatic arc

*Back-Arc Basins*
>    If subduction rate > rate of movement, overriding plate thins
>    Tension and thinning cause a sag, may have basaltic magma

*Transform Fault Margins*
>    Hugh, vertical, strike-slip faults, intense shattering
>    Numerous shallow focus earthquakes, some of high magnitude
>    Best-known one in No. Am. is the San Andreas in California

## Hot Spots and Absolute Motions

>    Independent reference needed to determine absolute motions
>    Hawaiian Islands appear to have formed over long-lived hot spots
>    100 hot spots have been described
>    Africa appears to be stationary and Atlantic Ridge moving west
>    Various plates exhibit growth/destruction characteristics

## Cause of Plate Tectonics

>    Exact cause remains to be agreed upon
>    Relative importance of lithosphere - asthenosphere activity in ??
>    1. lithosphere must have kinetic energy in order to move
>    2. source of kinetic energy is the Earth's internal heat
>    Heat is transfered through convection motion

*Convection in the Mantle*

Even though the mantle is solid rock, it appears to strain slowly

Convective motion is only one of several causes of plate motion

*Movement of the Lithosphere*

Three forces appear to be involved in plate motions

1. push away from a spreading center, creation of new seafloor
2. dragging, *i.e.*, descending tongue of old, cold lithosphere
3. slide downhill from spreading centers

Lithosphere is too weak and brittle to move by any one of these

Calculations suggest all 3 operate at different times and places

# Plate Tectonics, Continental Crust, and Mountain Building

Continental crust is sort of a passenger on the lithosphere

The "passenger" is buffeted, stretched, fractured, and altered

Continental crust formed by bump-and-grind tectonics

All history before mid-Jurassic is from the continents

*Regional Structures of Continents*

*Cratons*

On a continental scale 2 levels of structural units

Craton is the core of tectonically, isostatically stable ancient rock

*Orogens*

Elongate regions of intensely folded, faulted rock around craton

Orogens differ in age, size, history, and details of origin

*Continental Shields*

An assemblage of cratons and orogens in isostatic equilibrium

Stable platform, continental shield with a thin layer of sediments

Fig.16.19 several ancient cratons and orogens in Canadian Shield

Canadian Shield has several Archeon Eon minicontinents

Plate tectonics operated at least 2 b.y. ago

*Continental Margins*

Five types of continental margins:

passive, convergent, collision, transform fault, accreted terrane

*Passive Continental Margins*

One that occurs in the stable interior of plate *i.e.*, Atlantic marg.

Places of accumulation of great thicknesses of sediment

Clastic non-marine clastics, evaporites, clastic marine sediments

Plate triple junction, where 3 spreading centers intersect

Two arms of a triple junction succeed, one arm fails

*Continental Convergent Margins*

Where the edge of a continent coincides with convergent margin

Andean coast of South America

Mélange is a common environment

2 metamorphic belts with andesitic volcanics/granites between

*Continental Collision Margins*

Where 2 continents, on two different plates, collide

Alpine - Himalayan collision zone

All sediment caught between 2 colliding blocks forms mountains

Modern examples are fold-and-thrust mountain systems

Fold-thrust systems develop from sed. strata 15,000+ m thick

Passive continental margins become continental convergent

Collision line marked by serpentinites

The new mountain system lies in interior of a major landmass

Appalachians, Urals, Himalayan mountain systems

*Transform Fault Margins*

When the margin of a continent coincides with a transform fault

San Andreas Fault in California

Transform faults connect 2 (old) spreading centers

*Accreted Terrane Margins*

Former convergent or transform fault margin w/rafted-in exotics

Exotic blocks can be rafted in form great distances, 5,000 km

Each exotic fragment is a terrane

Numerous examples from western North America, <200m.y. old

Adds a second stage of complexity to plate tectonic theory

*Mountain Building*

Today's fold-and-thrust mountains, orogens<few hundred m.y.

*The Appalachians*

  Paleozoic fold-and-thrust, 2,500 km long

  Passive margin, continental shelf, shallow water deposits

  Thicken from west to east

  NY & PA west to east, change from flat lying to fold-and-thrust

  Further south(TN, NC, SC) great thrust faults(E to W), basement

  History covers 600 m.y.

*The Alps*

  Mesozoic - Cenozoic age deformation, south to north

  Shallow water sediments in north, deeper water in high Alps

*The Canadian Rockies*

  Structurally similar to the Appalachians

  Fold-and-thrust, and metamorphism

## Plate Tectonics and Climate: When Australia was Wet

  Western Australia is a continental shield >2.5 b.y., 100 to 500 m

  Underlying lithology controls topography

  200 mm precipitation/yr, poorly developed drainage, playas, etc.

  The shield is covered by weathering to hundreds of meters deep

  1. shield has been exposed for a very long time

  2. climate was much wetter sometime in the past

  Some laterite soils exist, typical of high rainfall, 30 to 200 m.y.

  Presently at 20° to 35°, absolute motion is about 5 cm/yr

   40 m.y. ago W. Australia was 2,000 km south, temperate climate

## OVERVIEW

This chapter basically of three parts: historical development of the idea of plate tectonics, causes of plate tectonics, and results of plate tectonics.

The section on development of the concept is any excellent opportunity to demonstrate how scientific thought evolves and to demonstrate that scientists are not so different that other people and can harbor deep biases that are hard to over come.

The section on causes of plate movements has more conjecture than most geologists would like. to go with the facts. Again this is a good opportunity to demonstrate that scientists must sometimes function with limited data.

The results of plate tectonics leaves the chapter on a good solid footing of observable facts. This is a fun chapter, enjoy yourself while you lecture it!

## LEARNING OBJECTIVES

1. Know that the lithosphere is broken into six large and many smaller plates.

2. Know that lithospheric plates are about 100 km thick.

3. Know that each lithospheric plate slowly moves over the top of the weak asthenosphere.

4. Know that there are three types of lithospheric plate margins: Divergent, Convergent, and Transform fault.

5. Know that divergent plate margins (spreading centers) are those where new lithosphere forms; plates move away from them.

6. Know that convergent plate margins (subduction zones) are lines along which plates compress each other and along which lithosphere capped by oceanic crust is subducted back into the mantle.

7. Know that transform fault margins are lines where two plates slide past each other.

8. Know that plate movement can be described in terms of rotation across the surface of a sphere.

9. Know that each plate rotates around a spreading axis that does not necessarily coincide with the Earth's axis of rotation.

10. Know that because plate movement is a rotation, the velocity varies from place to place on the plate.

11. Know that each segment of oceanic ridge that marks a divergent margin between two plates lies on a line of longitude passing through the spreading pole.

12. Know that each transform fault margin between two plates lies on the line of latitude of the spreading pole.

13. Know that the mechanism that drives a moving plate is not known, but apparently it results from a combination of convection in the mantle plus forces that act on a plate of the lithosphere.

14. Know that two major structural units can be discerned in the continental crust: cratons and separating orogens.

15. Cratons are ancient portions of the crust that are tectonically and isostatically stable.

16. Know that cratons are separated and surrounded by orogens of highly deformed rock, marking the site of mountain ranges.

17. Know that a continental shield is an assemblage of cratons and deeply eroded orogens that forms the core of a continent.

18. Know that there are five kinds of continental margins: passive, convergent, collision, transform fault, and accreted terrane.

19. Know that passive margins develop by rifting of the continental crust. The Red Sea is an example of a young rift, the Atlantic Ocean is a mature rift.

20. Know that a characteristic sequence of sediments forms along a passive continental margin, starting with clastic nonmarine sediments, followed by marine evaporites, and then marine clastic sediments.

21. Know that continental convergent margins are the locale of paired metamorphic belts, chains of stratovolcanoes (magmatic arc), and linear belts of granitic batholiths.

22. Know that collision margins are the locations of fold-and-thrust mountain systems.

23. Know that transform fault margins occur where the edge of a continent coincides with the transform fault boundary of a plate.

24. Know that accreted terrane margins arise from the addition of blocks of crust brought in by subduction and transform fault motions.

## LECTURE DEMONSTRATIONS

1. Good, high quality remote sensed images of continental margins give an excellent feel for plate margin activities.

2. Exercise care when making cross sections to keep them in proportion. Otherwise you can totally distort what is happening to the point of fabrication.

3. Pulling sheets of paper through the crack between two tables can be used to simulate seafloor spreading.

## FILMS AND VIDEO CASSETTES

1. <u>Continental Drift: The Theory of Plate Tectonics,</u> 22 min., 1980, (source: EBE).

2. <u>Continents Adrift,</u> 16 min., 1982, (source: EBE).

3. <u>The Drifting Continents, 50 min., 1971,</u> (source: TIM, TLV).

4. <u>Drifting Continents,</u> 24 min., 1979, (source: UM).

5. <u>Histoire d'une Baic,</u> 19 min., 1980, (source: FACSEA).

6. <u>How Solid is Rock?,</u> 21 min., 1967, (source: EB, BU, IU, USC).

7. <u>The Not so Solid Earth,</u> 30 min., 1971, (source: TLV).

8. <u>Planet Earth,</u> (source: FI).

9. <u>The Restless Earth,</u> Part I: <u>Plate Tectonic Theory,</u> 58 min., 1972, (source: IU).

10. <u>Spreading Oceans,</u> 24 min., 1978, (source: UM).

## REFERENCES

Bird, J. M., and J. F. Dewey, 1970, Lithosphere Plate - Continental Margin Tectonics and the Evolution of the Appalachian Orogeny, <u>Geo;logical Society of America Bulletin,</u> v. 81, pp. 1031-1060.

Bird, J. M., and B. Isacks, 1972, <u>Plate Tectonics: Selected Papers from the Journal of Geophysical Research,</u> Washington, D. C.: American Geophysical Union.

Carey, S. W., 1988, <u>Theories of the Earth and Universe: A History of Dogma in the Earth Sciences,</u> Stanford, CA: Stanford Univ. Press.

Chamberlain, T. C., 1897, The Method of Multiple Working Hypotheses, <u>Journal of Geology,</u> v. 5, pp. 837-848 (reprinted: <u>Science,</u> v. 148, pp. 754-759). **HAVE YOUR STUDENTS READ THIS CLASSIC PAPER.**

Condie, K. C., 1976, <u>Plate Tectonics and Crustal Evolution,</u> New York: Pergamon Press.

<u>Continents Adrift: Readings from Scientific American,</u> 1972, San Francisco: Freeman.

<u>Continents Adrift and Continents Aground: Readings from Scientific American,</u> 1976, San Francisco: Freeman.

Cox, A., ed., 1973, Plate Tectonics and Geomagnetic Reversals, San Francisco: Freeman.

Cox, A., and R. B. Hart, 1986, Plate Tectonics, Palo Alto, CA: Blackwell Scientific Publ.

Davies. P. A., and S. K. Runcorn (eds.), 1980, Mechanisms of Continental Drift and Plate Tectonics, New York: Academic Press.

LePinchon, X., J. Francheteau, and J. Bonnin, 1973, Plate Tectonics, New York: Elsevier.

Marvin, U. B., 1973, Continental Drift: The Evolution of a Concept, Washington, D. C.: Smithsonian Inst. Press.

Meyerhoff, A. A., and H. A. Meyerhoff, 1972, The New Global Tectonics: Major Inconsistencies, American Association of Petroleum Geologists Bulletin, v. 56, pp. 269-336.

Miller, R., 1983, Continents in Collision, Alexandria, VA: Time-Life Books.

Sullivan, W., 1974, Continents in Motion, New York: McGraw-Hill.

Talwani, M., and W. C. Pittman III (eds.), 1977, Island Arcs, Deep Sea Trenches and Back-Arc Basins, Washington, D. C.: American Geophysical Union.

Uyeda, S., and H. Kanaamori, 1979, Back-Arc Opening and the Mode of Subduction, Journal of Geophysical Research, v. 84, pp. 886-898.

Watkins, J. S., L. Montadert, and P. W. Dickerson (eds.), 1979, Geological and Geophysical Investigations of Continental Margins, Memoir 29, Tulsa, OK: American association of Petroleum Geologists.

Wegener, A., 1929, The Origin of Continents and Oceans, New York: Dover.

Wyllie, P. J., 1976, The Way the Earth Works, New York, Wiley.

# CHAPTER 17

## RESOURCES OF MINERALS AND ENERGY

### OUTLINE

### *Vignette on civilization and natural resources*

       Natural resources make civilization as we know it

       Early use was homogeneous rocks for tools, salt for diet

       Metals entered 17,000 yr. ago, copper and gold

       6,000 yr. ago copper smelting, iron only 3,500 yr ago

       Oil first used for fuel 4,500 yr. ago

       2,500 years ago various non-metallics mined; cement, glass

### Mineral Resources

       Modern civilization operates with machines

       Machines require natural resources to build and supply energy

       Mineral and energy resources sites are restricted and unusual

       Modern society requires a variety of metallics and nonmetallics

*Supplies of Minerals*

       Many nations have a great variety of mineral resources

       No country has all of the minerals to function, trade is needed

       Three aspects of minerals:

       1. Usable mineral deposits are restricted and localized

       2. total quantities of a mineral are not accurately known

       3. quantities are finite and nonrenewable

       Skills of exploration geologists to find and engineers to recover

*Ore*

       A profitable mineral deposit

       Whether or not a deposit is ore or is controlled by economics

*Gangue*

    Nonvaluable minerals mixed with the valuable one(s)

    Ore challenge is twofold:

    1. find the ore

    2. mine ore and dispose of gangue as cheaply as possible

# Origin of Mineral Deposits

    Ore deposits exist due to local enrichment of 1(+) minerals

    Mineral deposits are classified by the process of concentrated

    Five concentration processes:

    1. hydrothermal mineral deposits-hot, aqueous solutions

    2. magmatic mineral deposits-processes within a magma

    3. sedimentary mineral deposits-precipitation for a water body

    4. placer-concentration by running water

    5. residual mineral deposits-weathering processes

*Hydrothermal Mineral Deposits*

    Many famous mines formed this way, most common process

    Fig. 5.20 - volcanogenic massive sulfide deposits

    Elemental constituents are leached out of crust by hot water

    Similar composition fluids may have very different histories

    Important question-Why did precipitation of minerals occur?

*Causes of Precipitation*

    If cooling is too slow, precipitation is spread over too much area

    Very rapid cooling(fractures, high porosity) yields ores

*Examples of Precipitation*

    Veins form due to precipitation in fractures

    May be related to volcanic or magmatic cooling activities

*Magmatic Mineral Deposits*

    Fractional crystallization, pegmatites, variety of minerals

    Crystal settling, heavy minerals in low-viscosity magmas

*Sedimentary Mineral Deposits*

    Precipitation of substances carried in aqueous solution

*Evaporite Deposits*

    Layers of salt precipitated from evaporating water bodies

    Large variety of (hydrated) minerals, desert environments

    Lakes and the ocean precipitate different suites of minerals

*Iron Deposits*

        Iron deposits are widespread

        Lake Superior-type iron deposits

        Formed from oceans the were probably different from today's

        Lake Superior-type iron deposits are not ores, must weather

        Lake Superior-type iron deposits must be metamorphosed

*Stratabound Deposits*

        Thinly, interbedded sulfide-sediment deposits, diagenetic

        Form when hydrothermal solutions alter sediments

        Source of some huge ore bodies

*Placers*

        High specific gravity minerals concentrated by water currents

        More than half of all gold ever received has been placer

        Both modern and ancient gold placer deposits

        Witwatersrand Basin placer ores are running out

        Most new gold discoveries are hydrothermal

*Residual Mineral Deposits*

        Chemical weathering remove more soluble minerals

        Least soluble minerals are concentrated as residual deposits

        Laterites (Fe and Al) are a good example of residual minerals

        Most Bauxite is >90% formed in last 60 m.y., large are <25 m.y.

*Metallogenic Provinces*

        Limited region of crust w/unusually high concentration deposits

        Metallogenic provinces, weathering or plate tectonic controlled

## Useful Mineral Substances

        Mineral products are grouped on basis of use rather than origin

        1. metal yielding minerals

        2. nonmetallic minerals

        Metallic minerals can be mined only where concentrated

*Geochemically Abundant Metals*

        ≥0.1% by weight of the crust

        Require comparatively little enrichment-oxides and hydroxides

        Mostly residual, sedimentary, and magmatic deposits

*Geochemically Scarce Metals*

  <0.1% by weight of the crust

  Generally not found in common rocks

  Presence is usually the result of ionic substitution in minerals

  Hydrothermally concentrated

  Commonly occur as sulfides, oxides, or elemental form

  Mostly hydrothermal, magmatic, or placer

# Energy Resources

  North Americans consume large amounts of energy daily

  Consumption: transportation, home, commerce, industry

  Huge amounts of energy consumed annually

  Coal, oil, natural gas, nuclear, wood, dung, solar, hydroelectric

  Equals 2 metric tons coal or 10 barrels of oil/ each person/year

  Utilization is not uniform worldwide

*Supplies of Energy*

  Fossil fuels, hydroelectric, nuclear power, wood, wind, muscle

*Fossil Fuels*

  Remains of plants or animals trapped in sediments, used for fuel

  In oceans - microphytoplankton and bacteria trapped in shales

  On land - trees, bushes, grasses trapped in shales

  Oil shale, containing kerogen, forms if shale is not hot enough

*Coal*

  The most abundant fossil fuel

  Mostly used to produce steam to produce electricity or for coke

  Also an important raw material in the chemical industry

  Coalification - loss of volatiles, water, carbon dioxide, methane

  Coalification - peat, lignite, (sub)bituminous, anthracite

  Grade goes up, fewer volatiles make harder to ignite, less smoke

  Extreme metamorphism of coal produces graphite

*Occurrence of Coal*

  Swamp deposit, flat lens shaped body, slowly subsiding basin

  Two predominant types of basins:

  1. slowly subsiding basins in continental interiors

  2. swampy margins, shallow inland seas at times of high sea level

*Coal-Forming Periods*
> Tropical or subtropical climate
> Coal swamps occur throughout geologic history since Silurian
> 1. Pennsylvanian and Permian on Pangaea
> 2. Mid-Jurassic thru mid-Tertiary, peak in Cretaceous

*Petroleum: Oil and Natural Gas*
> Major oil utilization began in 1847 with marketing for lubricant
> 1852 production of kerosene
> 1856 hand dug wells in Rumania
> 1859 first drilled well at Titusville, PA, then "exploded" as a fuel
> Natural gas first used in China about 1000 B.C.
> 17th century in Europe, wood and coal gas for illumination
> Commercial gas companies, London, 1812, and Baltimore, 1816

*Origin of Petroleum*
> Decomposition of organic material trapped in sediment
> Migrates thru aquifers and is trapped in other sediments
> Petroleum forms in one rock, is recovered from a different one
> Oil migrates up slope in quartz and carbonate aquifers
> Trapped under some sort of impermeable cap
> Most oil escapes and is never trapped
> Most recovered oil is Cenozoic, has not been lost

*Distribution of Oil*
> Petroleum is not uniformly distributed in time or space
> Organic must be converted to petroleum, then be trapped
> Petroleum forms at a geothermal gradient of 1.8 to 5.5 C°/100 m
> Approximately 600 billion barrels already pumped from ground
> 1500 to 3000 billion barrels of oil may eventually be discovered

*Tars*
> Oil that is exceedingly viscous and will not flow easily
> Most be mined and heated to recover the oil

*Oil Shale*
> Kerogen trapped in shale
> Must be mined and heated to recover the oil
> Must yield >40 liters/ton to be economical
> World's largest deposit, Green River Fm., Eocene, UT, CO, WY
> Currently oil shale isn't exploited, must wait for better economic

*How Much Fossil Fuel?*

Are supplies of fossil fuels adequate to meet future demands?

0.22 tons of coal produces as much energy as 1 barrel of oil

$13,800 \times 10^9$ tons of coal reserves = $63,000 \times 10^9$ bbl of oil

The only long term source of energy seems to be coal

# Other Sources of Energy

Three other sources well developed

1. biomass - plant material
2. hydroelectric
3. nuclear energy

Five more await further development

1. solar radiation
2. wind
3. waves
4. tides
5. Earth's internal heat

*Biomass Energy*

Approximately 14% of world's energy needs

Most widely used in developing nations

Better management of vegetative resources might increase use

*Hydroelectric Power*

Recovered from potential energy of falling water

Sites to construct facilities are very limited

Reservoirs have finite life expectancy, "quickly" silt up

*Nuclear Energy*

Heat energy produced during nuclear fission

Appropriate isotopes are relatively rare

Only the most rare uranium isotope will sustain atomic reactions

The heat that is produced is used to produce steam for turbines

Nuclear power, most extensively used where fossil fuels are lack

*Geothermal Power*

Earth's internal energy being emitted

Generally utilized as dry steam to operate turbines

*Energy from Winds, Waves, Tides, and Sunlight*

> Photovoltaic cells to convert solar to electric energy, inefficient
>
> Wind is locally significant due to irregularities of wind
>
> Waves are not yielding much because harnessing devices corrode
>
> Tides are very inefficient sources of energy

## Mineral Deposits Forming Today

> 1962, accidental discovery of 320°C brine well in Imperial Valley
>
> Loaded with metal ions
>
> 1964, brine pools on floor of Red Sea, rising on normal faults
>
> 1978, 300°C springs with sulfide deposits on East Pacific Rise

# OVERVIEW

This chapter is divided into two broad section. The first covers mineral resources and their origin. The second part covers energy resources and their origin.

This is an extremely important chapter for students to comprehend. Few people seem to understand that non-renewable mineral and energy resources are absolutely finite. They should also understand the difference between reserves - recoverable with current technologies and economics, and resources - in the ground, potentially recoverable.

In the section on minerals the students need to understand that some outside influence, geological and/or biological is responsible for the formation of mineral deposits. They also need to understand that the processes operate incredibly slowly by human standards and are extremely localized.

The section on energy resources gives an opportunity to give the students an understanding of the finite nature of most energy resources. What is in the ground is not the same as what is recoverable.

As geology professors there is a need to explain the civilization as we know it depends on the recovery of natural resources. When a country runs out of natural resources, or can no longer recover what it has, it is on the road the history, not the road to greatness. No nation is self sufficient in all resources and trade is necessary. Neither total devastation of the lands, nor total preservation of the lands, protection from development of natural resources, is realistic.

# LEARNING OBJECTIVES

1. Know that when a mineral deposit can be mined profitably, it is called an ore.

2. Know that the waste material mixed with ore minerals is gangue.

3. Know that mineral deposits form when minerals become concentrated in one of five ways: (1) precipitation from hydrothermal solutions to form hydrothermal mineral deposits; (2) concentration through crystallization to form magmatic mineral deposits; (3) concentration from lake water or seawater to form sedimentary mineral deposits; (4) concentration in flowing water to form placers; (5) concentration through weathering to form residual deposits.

4. Know that hydrothermal solutions are brines, and they can either be given off by cooling magma or else form when either groundwater or seawater penetrates the crust, becomes heated, and reacts with the enclosing rocks.

5. Know that hydrothermal mineral deposits form when hydrothermal solutions deposit dissolved minerals because of cooling, boiling, pressure drop, mixing with colder water, or through chemical reactions with enclosing rocks.

6. Know that chromite, the main ore mineral of chromium, is the most important mineral concentrated by fractional crystallization.

7. Know that sedimentary mineral deposits are varied.

8. Know that the largest and most important sedimentary mineral deposits are evaporites.

9. Knot that marine evaporite deposits supply most of the world's gypsum, halite, and potassium minerals.

10. Know that gold, platinum, cassiterite, diamonds, and other minerals are commonly found mechanically concentrated in placers.

11. know that bauxite, the main ore of aluminum, is the most important kind of residual mineral deposit.

12. Know that bauxite forms as a result of tropical weathering.

13. know that the distribution of many kinds of mineral deposits is controlled by plate tectonics because most magmas and most sedimentary basins are where they are because of plate tectonics.

14. Know that geochemically abundant metals, which make up $\geq 0.1\%$ of the crust, tend to form residual, sedimentary, or magmatic mineral deposits.

15. Know that geochemically abundant metal deposits contain enormous amounts of metals available for exploitation.

16. Know that deposits of geochemically scarce metals, present in the crust in amounts >0.1%, form mainly as hydrothermal, magmatic, and placer deposits.

17. Know that amounts of scarce metals available for exploitation are limited and geographically restricted.

18. Know that nonmetallic substances are used mainly as chemicals, fertilizers, building materials, ceramics, and abrasives.

19. Know that coal originated as plant matter in ancient swamps and is both abundant and widely distributed.

20. Know that oil and gas originated as organic matter trapped in shales and decomposed chemically due to heat and pressure following burial.

21. Know that after formation oil and gas formed moved as fluids through reservoir rocks and were caught in geologic traps to form pools.

22. Know that kerogen, solid organic matter in shales, will convert to oil and gas when heated.

23. Know that oil from shales is the world's largest resource of fossil fuel, but that it takes more energy to extract most of it than is recovered.

24. Know that nuclear energy is derived from atomic nuclei of radioactive isotopes, chiefly uranium.

25. Know that nuclear energy available from naturally occurring radioactive elements is the single largest energy resource now available.

26. Know that other sources of energy currently used to some extent are geothermal heat, energy from flowing streams, winds, waves, tides, and the Sun's heat.

## LECTURE DEMONSTRATIONS

1. Good quality hand samples of common ore minerals and fossil fuels, or slides of them, are helpful in class. Do not forget to tell the students what they are used for.

2. Video materials of mines and mining operations are always interesting, or if available a field trip to a mine is helpful. Good examples of reclamation projects are good for the students to see.

# FILMS AND VIDEO CASSETTES

1. Earth's Physical Resources, 50 min., 1985, (source: MG).

2. Energy From the Crust, 24 min., 1985, (source: MG).

3. Energy: A Matter of Choices, 22 min., 1974, (source: EB).

4. Geothermal Energy, 34 min., 1984, (source: MG).

5. Geothermal Power, 30 min., 1970, (source: UC).

6. Geothermal Energy from Dry Rock, 14 min., 1978, (source: LA).

7. Gifts from the Earth, 58 min., 1985, (source: FI).

8. Hard-Rock Gold Mining, 20 min., 1974, (source: UC).

9. Mining for Nickel, 30 min., , (source: INCO).

10. Oil From Small Beginnings, 25 min., 1981, (source: UM).

11. Riches From the Earth, 20 min., 1982, (source: NGS).

# REFERENCES

Agricola, G., 1556, *De Re Metallica*, English translation by H. C. Hoover and L. H. Hoover, 1950, New York: Dover.

Bowen, R., 1979, Geothermal Resources, New York: Halsted Press.

Brobst, D. A., and W. P. Pratt, 1973, United States Mineral Resources, U. S. Geological Survey Professional Paper 820.

Earney, F. C. F., 1980, Petroleum and Hard Minerals From the Sea, New York: Halsted Press.

Flawn, P. T., 1966, Mineral Resources, New York: Wiley.

Jensen, M. L. and A. M. Bateman, 1979, Economic Mineral Deposits, New York: Wiley.

Park, C. F., Jr., 1968, Affluence in Jeopardy, San Francisco: Freeman, Cooper and Co.

Park, C. F., Jr., 1975, Earthbound: Minerals, Energy, and Man's Future, San Francisco: Freeman, Cooper and Co.

Raymond, R., 1986, <u>Out of the Fiery Furnace.</u> University Park, PA: The Pennsylvania State University Press.

Sawkins, F. J., 1984, <u>Mineral Deposits in Relation to Plate Tectonics.</u> New York: Springer-Verlag

Shih, C. C., J. E. Cotter, C. H. Prien, and T. D. Nevens, 1979, <u>Technological Overview Reports for Eight Shale Oil Recovery Processes.</u> Washington, D. C.: U. S. Environmental Protection Agency, EPA 600/7

Skinner, B. J., 1969, <u>Earth Resources.</u> Englewood Cliffs, NJ: Prentice-Hall.

Tissot, B. P., and D. H. Welte, 1984, <u>Petroleum Formation and Occurrence.</u> New York: Springer-Verlag.

Wright, J. B., 1977, <u>Mineral Deposits, Continental Drift, and Plate Tectonics.</u> Stroudsburg, PA: Dowden, Hutchinson and Ross, Inc.

Yen, T. F., and G. V. Chilingarian (eds.), 1976, <u>Oil Shale.</u> New York: Elsevier

# CHAPTER 18

## BEYOND PLANET EARTH

### OUTLINE

***Vignette on planetary rings***

    Rings of Saturn seem to become more amazing and complex

    Five rings visible from Earth, but they are many smaller rings

    Rings composed of tiny fragments of varying compositions

    Origin of rings unknown, but 3 other Jovian planets have them

**The Solar System**

    Mercury, Venus, Mars, Jupiter, Saturn known since antiquity

    Uranus, Neptune, Pluto discovered by telescope

    1957, artificial satellites opened new phase of Planetology

*Planets and Moons*

    Four terrestrial planets, sp. gr. $> 3.0$ gm/cm$^3$

    Five jovian planets, sp. gr. $< 2.0$ gm/cm$^3$

    Terrestrial planets, small, rocky, similar, but each unique

    Mercury, Venus, Earth, Mars

    Jovian planets, large, icy, similar, but each unique

    Jupiter, Saturn, Uranus, Neptune, Pluto(small)

    Jovian planets seem to each have a rocky core w/thick ice layer

    Orbits elliptical, lie in same plane, the ecliptic

    Planets probably all formed at same time in a similar manner

*Meteoroids and Asteroids*

    Meteoroids are small stony or metallic objects orbiting the sun

    Titius-Bode rule, distance to planet about 2X as far as next inner

    Asteroids in a belt where another planet should lie, Mars/Jupiter

    Asteroids and meteoroids seems to be in the terrestrial realm

*Cratered Surfaces*

Earth surface shaped by: tectonics, magmatism, gradation, crater

Impact craters formed by an impacting body - a bolide

200+ ancient impact craters on earth, none observed forming

Fig. 18.1 - laboratory derived sequence of crater formation

Numerous postimpact events modify the crater

1. walls may slump

2. isostatic rebound may produce changes in floor and rim

3. erosion fill crater with debris

4. basaltic magma rise along fissures

Cratering more abundant on other planetary bodies, all sizes

Weathering tends to obliterate them on Earth

## The Terrestrial Planets

Volcanism is widespread on the terrestrial planets and moon

Other members it is predominantly basaltic

Each planet has gone, or is going, through partial basalt melting

Difference between rocky planets, size/distance not composition

*Mercury*

Innermost planet, closest to sum

4880 km diameter, density 5.4 gm/cm$^3$, no atmosphere

Metallic core about 3600 km dia., impact crater pockmarked

No sign of tectonic activity

Magnetic field 1/100 Earth's

*The Moon*

Stark, impact cratered, basaltic volcanism, size of small planet

moon is terrestrial, probably formed in inner solar system area

*Structure*

Moonquakes are rare, 400/year, but yield much information:

1. crust on side toward Earth is 65 km thick

2. Moon is layered

3. Layer of regolith a few to several tens of meters thick

4. below regolith is a layer of shattered, broken rock

5. below that is 23 km of basalt then 40 km feldspar-rich rock

6. mantle begins at 65 km depth

156

Moonquakes are gravitationally caused by Earth

Moonquakes to 1000 km depth, rigid to that depth

Moon's density of 3.3 gm/cm$^3$ indicates core is very small

## The Lunar Surface

Two areas - lunar highlands (older) and maria (impact craters)

Highlands are tens of thousands of meters high

Gravity indicates they have roots, isostatically balanced, float

## Rocks and Regolith

Oldest are feldspar-rich anorthosite, 4.5 b.y., 100 m.y. after fm.

2nd, basalt with high K and P conc., last of highland activity

Youngest, basalt with high Fe and Ti, 3.2 to 3.8 b.y.

Regolith covers surface, gray pulverized rock frags., small dust

## History of the Moon

Formed as a solid 4.6 b.y. ago, accreted innumerable bolides

Gravitational field grew with size and accreted more bolides

Outer 150 to 200 km melted due to impacts, magma ocean

Magma ocean crystallized within first 400 m.y., done 4 b.y. ago

Interior heated up, magmas leaked out 3.8 to 3.0 b.y. ago

Tectonically and magmatically dead since

## Lessons for the earth

Early in history Earth probably had a magma ocean, but thicker

All traces of the primitive crust have been destroyed

## Mars

6787 km diameter, 1/10 Earth's mass, days nearly same length

Atmosphere 1/100 dense as Earth, mostly carbon dioxide

Lacks dipolar magnetic field

Composition and structure of Earth and Mars must be similar

No seismic information about Mars

## Surface Features

Fig. 18.9 - shows Mars' extraordinary topography

So. hemisphere is densely cratered similar to Moon, Mercury

No. hemisphere less cratered, smoother, more volcanism

20+ huge shield volcanos, plus many smaller ones, huge size

1. long-lived source of magma in interior

2. magma source connected to surface for a very long time

3. lithosphere is very thick and strong, isostacy very slow
Volcanism is recent, <100 m.y., due to no impact craters
Valles Marineris, huge(4500 km) canyon system, giant grabens

*Erosion*

Wind driven sand is principle erosive agent, storms last months
Evidence of extensive water erosion, but no free water apparent

*Geological History*

Early history similar to that of Earth and the Moon
Enough water was released to form a layer 50 m to 100 m thick
Weathering a erosion probably locked up $H_2O$ and $CO_2$
Lowered temperature would lead to permafrost in ground

*Venus*

Most like Earth in size and mass
Atmosphere 100X as dense as Earth's, carbon dioxide
Surface temperature 500°C, closer to Sun plus greenhouse effect
**Analysis of Venus' is currently in progress, watch for results**

*Volcanism*

Dominant process on Venus, volcanic plains cover 80%
Volcanism dominantly basaltic, some evidence of rhyolitic

*Tectonics*

Surface arrears to be unimodal, suggesting no plate tectonics
Two linear mountain ranges formed by compression-extension
Venus and Earth are more alike than Earth and any other planet

*Comparison of the Terrestrial Planets*

All have:
1. metallic core
2. undergone partial melting to form basaltic magma
3. been modified by impact cratering

Difference arise from:
1. Size of planet, controls atmosphere and thermal character
2. Distance of planet of Sun, determines phase of water
3. Presence or absence of life

## The Jovian Planets and Their Moons

*Jupiter*

> Retain record of gases that other planets probably formed from
>
> Gases dominated by: hydrogen, helium, ammonia, methane
>
> 2X as large as all other planets combined
>
> Gives off 2X as much energy as it receives

*The Jovian Moons*

> Four moons ≥ Earth's moon in size
>
> Densities indicate they are rocky bodies
>
> Io, closest, highly colored yellows, oranges(sulfur compounds)
>
> Volcanically active, both basaltic and sulfur, no impact craters
>
> Heat to drive volcanism is gravitational, not radiometric
>
> Next 3 moons decrease in density outward as ice layer thickens
>
> Europa's surface is intricately fractured, gravitational forces
>
> Ganymede, much thicker ice, appears to have ice tectonics
>
> Callisto, most distant, thickest ice, most impact cratered

*Saturn and its Moons*

> Composition similar to Jupiter's, smaller
>
> Emits more energy than it receives
>
> Most striking feature are the rings
>
> Titan is larger than mercury, has a substantial toxic atmosphere
>
> 45% rock, 55% ice, ethylene and acetylene smog covering
>
> Oceans of liquid ethane and methane w/ continents of ice

## Is There Life Out There?

> Only other possible planet with life is Mars, because of water
>
> Mars may have had free water and an atmosphere once
>
> No direct evidence of other life, possible in other solar systems

## OVERVIEW

Exploration of the solar system during the last 34 years is one of the great accomplishments of the Space Age. The great progress since *Sputnik* to the *Voyager II* and now the Space Shuttle will stand as monuments to this period of history.

The chapter deals first with the solar system in general. It presents a lot of the facts and dimensions that have been revised recently. Then the chapter deals with the Terrestrial Planets and their character and them with the outer planets and their character.

The Solar System is always a topic of interest to students and lectures go over well, especially when accompanied by slides.

## LEARNING OBJECTIVES

1. Know that the 4 planets closest to the Sun (Mercury, Venus, Earth, Mars) are small, dense, rocky bodies.

2. Know that the outer planets (Jupiter, Saturn, Uranus, Neptune, Pluto) are generally large, low-density, icy bodies.

3. Know that the Moon has a layered structure probably formed by differentiation.

4, Know that the Moon probably has a small core surrounded by a thick mantle and is capped by a crust 65 km thick.

5. Know that on the Moon, magma was formed early but is no longer generated.

6. Know that the Moon is a magmatically dead planet.

7. Know that the highlands of the Moon are remnants of ancient crust built by magmatic differentiation more that 4 billion years ago.

8. Know that the maria (lunar lowlands) are vast basins created by the impacts of giant meteorites and later filled in by lava.

9. Know that each of the terrestrial planets went through a period in internal radioactive heating that led to generation of basaltic magma.

10. Know that the Earth and possibly Mars and Venus are still producing magma from radioactive heating.

11. Know that Mars and Venus both appear to be one-plate planets, not multiplate planets like the Earth.

12. Know that Mars seems to be magmatically active. Olympus Mons, a shield volcano on Mars, is the largest volcano yet found in the solar system.

13. Know that the principal eroding agent on Mars is wind-driven sand and dust.

14. Know that water or some other flowing liquid cut stream channels on Mars at some time in the past.

15. Know that Venus has about the same size and density as the Earth, but has a dense atmosphere of carbon dioxide and a surface temperature of about 500°C.

16. Know that none of the other terrestrial planets has a climate hospitable to human life.

17. know that the large outer planets, Jupiter, Saturn, Uranus, and Neptune, have thick atmospheres that obscure their rocky cores.

18. Know that Pluto, the outermost planet, is probably covered by ice.

19. Know that the moons of Jupiter have progressively thicker outer layers of ice the further away they are from Jupiter.

## LECTURE DEMONSTRATIONS

1. The size of the solar system relative to the size of the bodies that constitute it can be illustrated by analogy. If the Sun is represented by a tennis ball. then the planets would be about the size of small grains of sand at distances of 2. 4, 5.4, 8, 28, 51, 103, 161, and 212 meters from the "sun". This is a rather graphic demonstration of how emptiness there is in space.

2. Excellent "geological" maps of the Moon and Mars are available from the U. S. Geological Survey.

3. Photographs and slides of the planets and their moons and the Sun are available from the suppliers listed elsewhere in this supplement.

## FILMS AND VIDEO CASSETTES

Apollo 15 in the Mountains of the Moon, 28 min., 1971, (source: NASA).

Journey Through the Solar System, (source: IV).

Mariner - Mars '69, 21 min., 1971, (source: NASA).

Mars - The Search Begins, 32 min., 1973, (source: NASA).

Mars, the Viking Mission, 24 min., 1981, (source: UM).

Planetary Exploration and Understanding the Earth, (source: IV).

A Question of Life, 23 min., 1976, (source: NASA).

The Planets, 52 min., 1976, (source: TIM).

Tales from Other Worlds, 60 min., 1985, (source: FI).

Universe, 28 min., 1976, (source: NASA).

The September 1979 and January 1980 issues of the Journal of Geological Education, contain reviews of numerous related films.

# REFERENCES

Beatty, J. K., 1981, The New Solar System, New York: Cambridge Univ, Press.

Carr, M. H.(ed.), 1984, Geology of the Terrestrial Planets, Washington, D. C.: NASA.

Greeley, R., 1987, Planetary Landscapes, Boston: Allen and Unwin.

Guest, P., P. Butterworth, J. Murray, and W. O'Donnell, 1979, Planetary Geology, New York: Wiley.

Hartman, W. K., 1972, Moon and Planets, Belmont, CA: Wadsworth Publ. Co.

Hartman, W. K., R. J. Phillips, and G. J. Taylor, 1986, Origin of the Moon, Houston: Lunar and Planetary Science Institute.

King, E. A., 1976, Space Geology, New York: Wiley.

Wilhelms, D. E., 1987, The Geologic history of the Moon, U. S. geological Survey Professional Paper 1348.

# CHAPTER 19

## OUR CHANGING PLANET

### OUTLINE

***Vignette on ozone***

Ozone in the atmosphere forms a shield to ultraviolet radiation
1985, a huge hole was found in the ozone layer over Antarctica
Chlorine from chlorofluorocarbons (CFCs) breaks down ozone
CFCs are widely used industrial chemicals
Several governments have pledged to eliminate CFCs by 2000

**The Dynamic Earth**

The Earth is being changed by man as well as nature
The Industrial Revolution accelerated mans impact on Earth
At the same time the standard of living was advanced
Man is a major factor in the Earth environment

*The Earth as an Interactive, Dynamic System*

Any change in solid, liquid, gas, organic realm affects the others
Results of mans activities are not always predictable or foreseen

*Time Scale of Change*

Realization that Earth is dynamic has come slowly and not totally
Natural equilibrium changes of Earth generally occur slowly
3 exceptions: large bolides, major volcanic eruptions, man

**The Changing Atmosphere**

Man asks numerous questions about his Earth environment
Three theories:
1. How the solid Earth works is called Plate Tectonics.
2. Dealing with life since earliest times is Evolution.
3. Dealing with the atmosphere is Climatology.
Climatology is most difficult because of multidiscipline nature

Many interacting subsystems, many levels, many time scales

Geology contributes the dimension of time to climatology

Climate will continue to change, but how and why is the question

## The Carbon Cycle

Moving thru biosphere, atmosphere, lithosphere, hydrosphere

Carbon occurs in 4 reservoirs:

1. atmosphere as carbon dioxide

2. biosphere as organic compounds

3. hydrosphere as dissolved carbon dioxide

4. crust as calcium carbonate and decaying/buried organics

Biosphere and atmosphere turnover every 4.5 years

Hydrosphere and crust cycle on a longer term, millions of years

Hydrosphere goes to $CaCO_3$ shells, crust to fossil fuels, etc.

Mans activities in the cycle:

1. extract and burn fossil fuels put $CO_2$ into atmosphere

2. remove forests slows down storage rates into organics

Carbon dioxide content of atmosphere is increasing rapidly

## The Greenhouse Effect

Sun supplies energy to drive atmosphere that drives climate

Short wave length energy from the sun passes thru easily

Long wave length energy from Earth surfaces is trapped

Greenhouse effect makes Earth habitable, Mars and Venus not

## Greenhouse Gases

Dry air - nitrogen, 79%, oxygen, 20%, argon, 1%

Water vapor accounts for 80% of greenhouse effect

Other gases in order: $CO_2$, methane, nitrous oxide, ozone, CFCs

## Trends in Greenhouse Gas Concentrations

Concentration of each gas appears to be increasing

## Carbon dioxide

Measurements on Mauna Loa volcano, Hawaii show 2 things

1. Amount of $CO_2$ fluctuates in an annual rhythm

2. Since 1958 conc. has increased 315,00 to 355,000 ppbv

Is the rise in $CO_2$ unusual, and can it be explained?

During ice ages $CO_2$ content was 200,000 ppbv

During interglacial age $CO_2$ content was 280,000 ppbv

355,000 ppbv represents a major increase with no precedent

Increase related to accelerated fossil fuel burning, deforestation

*Methane*

Absorbs infrared radiation 25X more efficiently that $CO_2$

Concentration is increasing for 3 reasons:

1. biological activity related to rice cultivation

2. leaks in domestic and industrial gaslines

3. byproduct of digestion of domestic livestock, especially cattle

*Other Trace Gases*

CFCs are incredibly efficient infrared radiation absorbers

Ozone and nitrous oxide in the troposphere are IR absorbers

*Global Warming*

What do these increases in concentration portend to climate?

*Historical Temperature Trends*

Few instrument measurements before 1850, most since WW II

Most records are for land masses of Northern Hemisphere

All curves indicate the same thing, a long-term temperature rise

*Climate Models*

General circulation models provide good weather predictions

GCMs are greatly simplified and leave many things out

A set of conditions are specified for a particular run of the model

Grid spacing is wide because of data and cost of computer time

*Model Estimates of Greenhouse Warming*

Five models are used for most of the predictions

All predict global warming of 0.5 to 1.5°C, 0.5°C has been shown

Predicted changes not uniform, greatest in polar regions

How this increase will affect climate changes is not understood

Scientists seem to agree on three things:

1. human activities have increased greenhouse gas concentration

2. global mean surface air temperature increased 0.3 to 0.6°C

3. next 100 years increase about 0.3°C/decade

*Environmental Effects of Global Warning*

The total temp. difference between glacial and interglacial, 5°C

Many other side effects of these activities

*Global Precipitation Changes*

      Temp. increases will increase evap., precipt. will be uneven

      Tropics wetter, interior of continents warmer and dryer

*Increased Storminess*

      Warmer, wetter tropics, larger, more frequent hurricanes

*Melting (and Growing?) Glaciers*

      Low- and middle-latitude glaciers might melt, high-latitude grow

*Reduction of Sea Ice*

      Northern sea ice might melt and reduce earth's albedo

*Thawing of Frozen Ground*

      Effect ecosystems and man made projects

*Rise of Sea Level*

      Glacial melting will raise sea level, flood populated coasts

      An increase in hurricane frequency might be expected also

*Changes in the Hydrologic Cycle*

      Shifting precipitation patterns could effect runoff/groundwater

*Decomposition of Soil Organic Matter*

      Increased temp. will accelerate decomp. increase $CO_2$ release

*Breakdown of Gas Hydrates*

      Ice-like solids with gas molecules locked in structure of water

      10,000 billion metric tons carbon(2X all fossil fuels) tied up

      Release of this by melting of frozen ground will amplify problem

      No one knows exactly what will happen, lots of predictions

## The Past as a Key to the Future

      Ayer's Law, "Anything that did happen, can happen."

      Look at historical geology for clues

*Lessons from the Past*

      Pre-human sudden climatic changes give clues

      End of last glacial, Greenland ice core showed 7°C rise in 40 yr.

      Pliocene temp. 3 to 4°C warmer, sea level tens of meters higher

      Marine organism migrated poleward; soil profiles reflect warmth

*Modeling Past Global Changes*

      GCMs can be compared to past events to see if they can predict

      Modeling has been tried for last 18,000 yrs, 3,000 yr intervals

Boundary conditions are very different for each level

Study Fig. 19.12 for various levels

In general, the predictions and geological evidence are close

## Perspectives of Global Change

Greenhouse gases are increasing, temp will increase 2 to 4°C

Temp. increases will be nonuniform, volcanic interruptions

This is a brief interlude and will end when fossil fuels are gone

Eventually another glacial event will occur due to Earth's orbit

Mid-Cretaceous temp. warmer, life zones 5 to 15° poleward

Mid-Cretaceous sea level 100 to 200 m higher than today

Several factors: geography, ocean circ., atmospheric composition

GCM models suggest geography plus $CO_2$ content could do this

$CO_2$ was probably from degassing of magmas in upper crust

Middle Cretaceous had unusually high volcanic activity

$CO_2$ levels could have increased 8 to 20X present, temp. 10°C+

Great lava out pourings could have resulted from super hot spots

# OVERVIEW

This chapter is a review of event that are transpiring to alter the atmosphere and a prediction of what may happen in the future. It give a feeling of helplessness that our environment is out of control and humans are being swept along with the environment.

This deals with items that are more likely to be read in recent publications and the popular media. Many news items will be found in that manner.

It may require conciderable effort to keep these lectures "on track" and not let the students and enthusasm get away from you.

## LEARNING OBJECTIVES

1. Know that synthetic chlorofluorocarbon (CFC) gases entering the upper atmosphere break down to chlorine, which destroys the protective ozone layer.

2. Know that discovery of a vast and recurring hole in the ozone layer over Antarctica has led to international efforts to eliminate CFC production by the end of this century.

3. Know that the Earth's climate system involves the atmosphere, hydrosphere, lithosphere and biosphere.

4. Know that changes affecting the climate system operate on time scales ranging from decades to millions of years.

5. Know that by using information from the geologic record, geologists can measure the magnitude and geographic extent of past climatic changes, determine the range of climatic variability on different time scales, and test the accuracy of computer models that simulate past climatic conditions.

6. Know that the carbon cycle is among the most important of the Earth's biogeochemical cycles.

7. Know that carbon resides in the atmosphere, the biosphere, the hydrosphere, and in the crust and cycles through these reservoirs at different rates.

8. Know that the anthropogenic extraction and burning of fossil fuels perturbs the natural carbon cycle.

9. Know that the anthropogenic extraction and burning of fossil fuels has led to an increase in atmospheric $CO_2$ since the start of the Industrial Revolution.

10. Know that the greenhouse effect, caused by the trapping of long-wave infrared radiation by trace gases in the atmosphere, makes the earth a habitable planet.

11. Know that the increase in the atmospheric trace gases ($CO_2$, $CH_4$, $O_3$, $N_2O$, and the CFCs) due to human activities is projected to warm the lower atmosphere by 2° to 4°C by the end of the next century.

12. Know that a probable 0.5°C increase in average global temperature since the mid-nineteenth century may reflect the initial part of this warming.

13. Know that the rate of warming is likely to reach 0.3°C per decade and may lead to a "super interglacial", making the Earth warmer than at any time in human history.

14. Know that potential physical and biological consequences of global warming include global changes in precipitation and vegetation patterns, increased storminess, melting of glaciers, sea ice, and frozen ground, a worldwide rise of sea level. local and regional changes in the hydrologic cycle, and increased rates of organic decomposition in soils.

15. Know that evidence of past intervals of rapid environmental change in the geologic record and reconstructions of past warmer intervals can provide insights into physical and biological responses to global warming.

16. Know that reconstructions of past episodes of climatic change permit evaluation of climate model simulations by general circulation models.

17. Know that although the Earrth's surface environments may change aubstantially during the next several centuries in response to greenhouse warming, viewed from the geologic perspective, this interval will appear as only a brief perturbation in the earth's climatic history.

## LECTURE DEMONSTRATIONS

1.  In light of the relatively complex nature of the topic of this chapter I suggest a computer program that may give some insights to the changes transpiring.

> SimEarth from MAXIS
> Broderbund Software, 17 Paul Drive, San Rafael, CA 94903-2101
> $69.95

## SOURCES OF INFORMATION

Carbon Dioxide Information Analysis Center, Oak Ridge National Laboratory, Building 1000, P.O.Box 2008, Oak Ridge, TN 37831-6335.

National Institute for Global Environmental Change(NIGEC), University of California at Davis, Davis, CA 95616

# SOURCES FOR 35 MM SLIDES AND RELATED MATERIALS

American Association of Petroleum Geologists, P.O.Box 979, Tulsa, OK 74101

Crystal Productions, P.O.Box 12317, Aspen, CO 81612

Educational Images Ltd., P.O.Box 3456, West Side, Elmira, NY 14905

Educational Materials and Equipment Co., Old Mill Plain Rd., Danbury, CT 06811

Geological Education Aids, 17 Leisure Drive, Kirksville, MO 63501

Geo-Tech Imagery, P.O.Box 3216, San Clemente, CA 92672

GeoPhoto Publishing Co., P.O.Box 1960, Orem, UT 84057

Geoscience Resources, 2990 Anthony Road, Burlington, NC 27215

Gould Media Inc., 44 Parkway West, Mount Vernon, NY 10552

Hubbard Scientific, P.O.Box 104, Northfield, Il 60065

JLM Visuals, 920 7th Ave., Grafton, WI 53024

McGraw-Hill Book Co., 330 West 42nd Street, New York, NY 10036
*Slides for Geology* by David Rahm

Michael Jay Publications, P.O.Box 1565, Loomis, CA 95650

Michael and Rosario Douglas, P.O.Box 13753, Boulder, CO 80308

MMI Corp., 2950 Wyman Parkway, P.O.Box 19907, Baltimore, MD 21211

National Geophysical Data Center, NOAA, Code E/GC4, 325 Broadway, Boulder, CO 80303

James L. Ruhle and Associates, P.O.Box 4301, Fullerton, CA 92634

Soil Science Society of America, 677 South Segoe Rd., Madison, WI 53711

Technology Application Center, University of New Mexico, Albuquerque, NM 87131

W. H. Freeman, 660 Market St., San Francisco, CA 94104 *Earth Science Slides* by John S. Sheldon, and *Physical Geology* by Warren Hamilton

Wards Natural History Establishment, P.O.Box 1712, Rochester, NY 14603 or P.O.Box 1749, Monterey, CA 93940

# SOURCES FOR FILMS AND VIDEOS

ABP     Arthur Barr Productions, 1029 North Allen Ave., Pasadena, CA 91004

AEF     American Educational Films, Box 5001, 132 Lusky Dr., Beverly Hills, CA 90212, (213) 278-4996

AIMS     AIMS Media, Inc., 626 Justin Ave., Glendale, CA 91201

AIS     Australian Information Service, Australian Consulate-General, 636 Fifth Ave., New York, NY 10111

AVP     Ambrose Video Production, Inc., 381 Park Ave. So., Suite 1601, New York, NY 10016

BBC     British Broadcasting Co. - Television, 630 Fifth Ave., New York, NY 10020

BF     Barr Films, P.O.Box 7878, Irwindale, CA 91706

BM     Bureau of Mines Audiovisual Library, Cochrans Mill Road, P.O.Box 18070, Pittsburgh, PA 15236

BU     Krasker Memorial Film Library, Boston Univ., School of Education, 765 Commonwealth Ave., Boston, MA 02215

CES     Charles Eames Studio, 901 Washington, Blvd., Venice, CA 90291

CFI     Canadian Film Institute, 1762 Carling Ave., Ottowa, Ontario K2A 2H7,CANADA

CFV     Coronet Feature Video, Simon and Schuster School Group, 108 Wilmot Road, Deerfield, Il 60015

CO     Circle Oaks Production, 73 Girdle Drive, Katonah, NY 10536

CRM     McGraw-Hill Films, 110 15th St., Del Mar, CA 92014

CT     California Institute of Technology, TV and Radio News Bureau, Pasadena, CA 91102

EB     Encyclopedia Britannica Films, Preview/Rental Library, 1822 Pickwick Ave., Glenview, IL 60025 (eastern and midwestern)

| | |
|---|---|
| EBE | Encyclopedia Britannica, Rental Library, 425 North Michigan Ave.,Dept. 10 PIR, Chicago, IL 60611 (312) 231-7331 |
| EM | Educational Media Center, The University of Colorado, Campus Box 379, Boulder, CO 80309 |
| ES | Earth Science Films, Canada, 2846 Athol St., Regina, Saskatchewan, Canada S4S 1Y2 |
| FA | Film Australia, Australian Information Service, Australian Consulate-General, 636 Fifth Ave., New York, NY 10111 |
| FACSEA | French American Culture Services and Educational Aid, 972 Fifth Ave., New York, NY 10021 (212) 370-4440 |
| FHS | Films for the Humanities and Sciences, P.O.Box 2053, Princeton, NJ 08543 |
| FI | Films Incorporated, 5547 N. Ravenswood Ave., Chicago, IL 60640 |
| GIS | Geoscience Information Services, P.O.Box 3326, Reston, VA 22090 |
| HUB | Hubbard, P.O.Box 104, Northbrook, IL 60065 |
| HUM | Humble Oil and Refining Co., PR/Advertising Dept., Box 2180, Houston, TX 77258 |
| INCO | INCO United States, Inc., One New York Plaza, New York, NY 10004 |
| ISU | Iowa State University, Media Resource Center, Film/Video Library, 121 Pearson Hall, Ames, IA 50011 |
| IU | Audio-Visual Center, Indiana University, Bloomington, IN 47401 |
| IV | Instructional Video, P.O.Box 21, Maumee, OH 43537 |
| JF | Journal Films, 930 Pitner Ave., Evanston, IL 60202 |
| JW | John Wiley and Sons, 605 Third Ave., New York, NY 10016 |
| KM | Karol Media, 625 From Rd., Paramus, NJ 07652 |
| LA | Los Alamos National Laboratory, Motion Picture/Video Production, Mail Stop D415, Los Alamos, NM 87545 |
| MC | Mar/Chuck Film Industries, P.O.Box 61, Mt. Prospect, IL 60056 |

| | |
|---|---|
| MG | The Media Guild, 11722 Sorrento Valley Rd., Suite E, San Diego, CA 92121 |
| MGH | McGraw-Hill Films, Princeton Rd, Hightstown, NJ 08520 (eastern states) |
| MGH | McGraw-Hill Films, 828 Custer Ave., Evanston, IL 60202 (midwestern states) |
| MGH | McGraw-Hill Films, 1714 Stockton St., San Francisco, CA 94133 (western states) |
| MLA | Modern Learning Aids, 1212 Avenue of the Americas, New York, NY 10036 |
| MMI | MMI Corp., P.O.Box 19907, Baltimore, MD 21211 |
| MP | Moonlight Productions, 2243 Old Middlefield Way, Mountain View, CA 94943 |
| MT | Modern Talking Pictures, 5000 Park St. N, St. Petersburg, FL 33709 |
| MTP | Modern Talking Picture Service, 2323 New Hyde Park, Hyde Park, NY 11040 |
| NA | National Archives and Records Administration, National Audiovisual Center, 8700 Edgeworth Drive, Capitol Heights, MD 20742 |
| NASA | NASA Headquarters Office of Public Affairs, Washington, D.C. (write for film list, which gives 9 regional addresses for free film borrowing) |
| NAVC | National Audiovisual Center, 8700 Edgeworth Dr., Capitol Heights, MD 20743 |
| NFBC | National Film Board of Canada, 1251 Avenue of the Americas, 16th Floor, New York, NY 10020 |
| NGS | National Geographic Society, Educational Services, Washington, D.C. 20036 |
| NWWA | National Water Well Assoc., P.O.Box 16737, Columbus, OH 43216 |
| OSU | Ohio State University, Film/Video Distribution Center, Dept. of Photography and Cinema, 156 West 19th Ave., Columbus, OH 43210 |

| OU | Open University Education Enterprises Ltd., 12 Cofferidge Close, Stony Stratford, Milton Keynes, England |
|---|---|
| PBS | PBS Video, 1320 Braddock Place, Alexandria, VA 22314 |
| PF | Pyramid Films and Video, P.O.Box 1048, Santa Monica, CA 90406 |
| PSU | Pennsylvania State University, Audio Services, Special Services Building, University Park, PA 16802 |
| PU | Purdue University, Audio Visual Center, Stewart Center, West Lafayette, IN 47907 |
| SOC | Shell Oil Co., 1433 Sadlier Circle, West Drive, Indianapolis, IN 46239 |
| TEC | Teledyne Exploration Co., P.O.Box 36269, 5825 Chimney Rock Rd., Houston, TX 77036 |
| TIM | Time-Life Education, Box 834, Radio City Post Office, New York, NY 10019 |
| TLV | Time-Life Video, P.O.Box 644, Paramus, NJ 07652 |
| TV | Tropical Visions, P.O.Box 1423, Keaau, HI 69749 |
| UC | University of California, Extension Media Center, 2176 Shattuck Ave., Berkley, CA 94704 |
| UM | University Media, 11526 Sorrento Valley Rd., Suite J, San Diego, CA 92121   (714) 755-9191 |
| USC | University of Southern California, Division of Cinema, Film Distribution Section, University Park, Los Angeles, CA 90007 |
| USFS | U. S. Forest Service, 1720 Peachtree Rd. NW, Atlanta, GA 30369 (404) 881-2384 |
| USGS | Chief, Visual Services, U. S. Geological Survey, 790 National Center, Reston, VA 22092 |
| USNAC | U. S. National AV Center, General Services Administration, Washington, D.C. 20409 |
| UW | Bureau of Audio Visual Instruction, University of Wisconsin - Extension, P.O.Box 2093, Madison, WI 53701 |

UWP        University of Washington Press, P.O.Box C-50096, Seattle, WA 98145

U T        University of Texas, Film Library, P.O.Box W, Austin, TX 78712

VS        The Volcano Show, Hellusundi 6A, Reykjavik, Iceland

# Part II

# TEST BANK

# CHAPTER 1

## PLANET EARTH

### MULTIPLE CHOICE

1. Nicolaus Copernicus was the first to _____ .

    *A - suggest that the planets revolve around the sun

    B - use a telescope to see mountains on the moon and moons of Jupiter.

    C - propose the concept of planetary accretion.

    D - propose that the earth is layered.

    E - forcefully propose the concept of Continental Drift.

2. Alfred Wegener was the first to _____ .

    A - suggest that the planets revolve around the sun

    B - use a telescope to see mountains on the moon and moons of Jupiter.

    C - propose the concept of planetary accretion.

    D - propose that the earth is layered.

    *E - forcefully propose the concept of Continental Drift.

3. Galileo Galilei was the first to _____ .

    A - suggest that the planets revolve around the sun

    *B - use a telescope to see mountains on the moon and moons of Jupiter.

    C - propose the concept of planetary accretion.

    D - propose that the earth is layered.

    E - forcefully propose the concept of Continental Drift.

4. The earth has an atmosphere composed dominantly of _____ .

    A - carbon, hydrogen, oxygen, and nitrogen

    B - nitrogen, oxygen, hydrogen, and water vapor

    C - nitrogen, carbon, water vapor, and carbon dioxide

    *D - nitrogen, oxygen, carbon dioxide, and water vapor

E - nitrogen, carbon dioxide, hydrogen, and oxygen

5. Regolith does not include _____ .

    A - soil        B - river mud        C - desert sand        D - friable rock debris

    *E - hard rock cliffs

6. Weathering is _____ .

    A - chemical alteration of rock        B - mechanical breakdown of rock

    C - the accumulation of mud, sand, and soil        D - all of the above

    *E - only A and B

7. Weathering is caused by _____ .

    A - exposure to atmosphere    B - exposure to biosphere    C - exposure to hydrosphere

    *D - all of the above        E - none of the above

8. The terrestrial planets are _____ .

    A - Mercury, Earth, Jupiter, Pluto    B - Jupiter, Saturn, Uranus, Neptune, and Pluto

    *C - Mercury, Venus, Earth, and Mars    D - Venus, Earth, Jupiter, Saturn, and Mars

    E - Mercury, Mars, Saturn, Uranus

9. The jovian planets are _____ .

    A - Mercury, Earth, Jupiter, Pluto    *B - Jupiter, Saturn, Uranus, Neptune, and Pluto

    C - Mercury, Venus, Earth, and Mars    D - Venus, Earth, Jupiter, Saturn, and Mars

    E - Mercury, Mars, Saturn, Uranus, Neptune

10. The terrestrial planets are _____ .

    *A - small, rocky, and more dense    B - small, massive, and less dense

    C - large, massive, and less dense    D - large, rocky, and less dense

    E - small, rocky and less dense

11. The jovian planets are _____ .

    A - small, rocky, and more dense    B - small, massive, and less dense

    *C - large, massive, and less dense    D - large, rocky, and less dense

    E - small, rocky and less dense

12. Planetary accretion is the process of _____ .

    A - the sun capturing planets from space    B - the planets capturing satellites

*C - the planets capturing small objects from space to grow larger

D - planets causing hydrogen and helium to condense

E - stars growing larger by capture of volatiles from space

13. Planetary accretion is caused by _____ .

    A - polar forces        B - electrical attraction      C - magnetic attraction

    *D - gravitational attraction      E - volatilization of gases

14. The probable causes for the earth undergoing partial melting are _____ .

    A - meteorite impacts        B - friction as earth revolved around the sun

    C - radioactive transformation of elements    D - all of the above   *E - only A and C

15. During partial melting of the earth volatiles such as _____ were released.

    *A - water vapor, carbon dioxide, and methane    B - water vapor, nitrogen, and argon

    C - nitrogen, oxygen, and water vapor    D - silicon, aluminum, sodium , potassium

    E - uranium, thorium, and potassium

16. Partial melting of the earth caused lighter materials high in _____ to rise to the surface.

    A - water vapor, carbon dioxide, and methane    B - water vapor, nitrogen, and argon

    C - nitrogen, oxygen, and water vapor    *D - silicon, aluminum, sodium , potassium

    E - uranium, thorium, and potassium

17. Partial melting of the earth caused the earth the alter from _____ .

    *A - homogeneous to compositionally layered    B - low density to high density

    C - high density to low density    D - compositionally layered to homogeneous

    E - caused no changes at all

18. The core and mantle are _____ .

    A - thinner than the crust    B - the same density   C - lower density than the crust

    D - far from uniform in thickness      *E - nearly uniform in thickness

19. The crust is _____ .

    A - thicker than the mantle        B - the same thickness as the core

    C - higher density than the mantle      *D - far from uniform in thickness

    E - nearly uniform in thickness

20. The oceanic crust is _____ than the continental crust.

A - thicker      *B - thinner      C - less uniform in thickness

D - less dense      E - less extensive

21. The continental crust is _____ than the oceanic crust.

     *A - thicker      B - thinner      C - less uniform in thickness

     D - less dense      E - less extensive

22. The crust and mantle _____ .

     A - are different in composition      B - have a distinct boundary

     C - are different in density      *D - all of the above      E - only A and C

23. Evidence that the core of the Earth is iron is suggested by _____ .

     A - the presence of iron meteorites      B - the high density of that portion of the Earth

     C - the Earth's magnetic field      *D - all of the above      E - none of the above

24. The portion of the earth that is composed of solid, high temperature iron.

     *A - inner core      B - outer core      C - mesosphere      D - asthenosphere

     E - lithosphere

25. The portion of the earth that is composed of liquid, high temperature iron.

     A - inner core      *B - outer core      C - mesosphere      D - asthenosphere

     E - lithosphere

26. The portion of the earth that is composed of solid, high temperature rock.

     A - inner core      B - outer core      *C - mesosphere      D - asthenosphere

     E - lithosphere

27. The portion of the earth that is composed of plastic, high temperature rock.

     A - inner core      B - outer core      C - mesosphere      *D - asthenosphere

     E - lithosphere

28. The portion of the earth that is composed of cooler, stronger, more rigid rock.

     A - inner core      B - outer core      C - mesosphere      D - asthenosphere

     *E - lithosphere

29. The lithospheric plates appear to move because of _____ .

     A - conduction in the asthenosphere      *B - convection in the asthenosphere

     C - conduction in the mesosphere      D - convection in the mesosphere

E - gravitational forces on the surface

30. The study of the movements and deformation of the lithosphere is _____ .

   A - plate tectonics     B - deformational dynamics     C - structural dynamics

   *D - tectonics          E - tectonophysics

31. The study of the lateral movements of the lithosphere over the asthenosphere is
_____ .

   *A - plate tectonics    B - deformational dynamics     C - structural dynamics

   D - tectonics           E - tectonophysics

32. The continents stand higher than the ocean basins because _____ .

   A - the ocean basins have been more deeply eroded than the continents

   B - the continents have had more deposition on them

   *C - the continental crust is lower density than the oceanic crust

   D - the continental crust is higher density than the oceanic crust

   E - the differences are not real, only apparent

33. The variable width margin of the continents that are flooded by ocean waters.

   *A - continental shelf     B - continental slope     C - continental rise

   D - continental crust      E - continental margin

34. The "steep" region landward of the geologic edge of the continent.

   A - continental shelf      *B - continental slope     C - continental rise

   D - continental crust      E - continental margin

35. The gently sloping portion of the ocean floor immediately oceanward of the edge of the
continent.

   A - continental shelf      B - continental slope      *C - continental rise

   D - continental crust      E - continental margin

36. The two most prominent features of the oceanic crust are _____ .

   A- ocean ridges and abyssal hills          B - abyssal hills and abyssal plains

   C - abyssal plains and trenches            *D - trenches and oceanic ridges

   E - trenches and abyssal plains

37. Along the center of the oceanic ridges _____ .

   A - rifts occur       B - intense volcanic activity occurs

C - a few volcanic islands occur        *D - all of the above        E - none of the above

38. Spreading centers occur at _____ .

   *A - divergent plate margins      B - convergent plate margins      C - subduction zones

   D - collision zones                E - transform fault margins

39. One plate goes under the other at _____ .

   A - divergent plate margins       B - convergent plate margins      C - subduction zones

   D - collision zones                *E - B and C

40. One plate slides past another at _____ .

   A - divergent plate margins       B - convergent plate margins      C - subduction zones

   D - collision zones                *E - transform fault margins

41. Near a spreading center, the crust is _____ .

   A - unusually thick       *B - unusually thin       C - cooler than usual

   D - no longer rigid       E - A and C

42. Further from a spreading center, the crust is _____ .

   A - thicker     *B - thinner        C - cooler        D - less rigid   E - A and C

43. Subduction zones are marked by _____ .

   A - abyssal plains        B - oceanic ridges        *C - deep trenches

   D - abyssal hills         E - sea mounts

44.  Collision zones are typically marked by _____ .

   A - deep trenches         B - hot spots   C - great vertical faults that move sideways

   *D - great mountain ranges         E - general quiescence

45.  Transform fault zones are typically marked by _____ .

   A - deep trenches         B - hot spots  *C - great vertical faults that move sideways

   D - great mountain ranges          E - general quiescence

46. The hydrologic cycle is powered by energy from _____ .

   A - hot springs        *B - the sun        C - gravity    D - geonuclear reactions

   E - plate tectonic activity

47. Rocks formed by the cooling and consolidation of magma.

   *A - igneous            B - sedimentary            C - metamorphic

48. Rocks formed from chemical precipitates or cementation of particles that have been transported and deposited by water, wind, or ice.

   A - igneous          *B - sedimentary          C - metamorphic

49. Rocks changed from the original form by high temperature and/or high pressure.

   A - igneous          B - sedimentary          *C - metamorphic

50.  Most of the surface of the continents is covered by _____ rocks.

   A - igneous          *B - sedimentary          C - metamorphic

51. Most of the volume of the continents composed of _____ rocks.

   A - igneous     B - sedimentary     C - metamorphic     D - A and B     *E - A and C

52. The most successful deep drilling project in the continental crust has been by the
_____ .

   A - U.S.A.     B - Germany     C - France     *D - U.S.S.R.     E - Japan

53. The most successful deep drilling project in the oceanic crust has been by the _____
.

   *A - U.S.A.     B - Germany     C - France     D - U.S.S.R.     E - Japan

## TRUE-FALSE

1. Study of seismic waves suggests that the density of the Earth increases at a uniform rate with depth.

   T/*F

2. The continents are still moving.

   *T/F

3. Modern shorelines nearly coincide with the boundaries between continental crust and oceanic crust.

   T/*F

4. All continental margins coincide with the edges of plates.

   T/*F

5. The boundaries of the lithospheric plate have been determined largely from the study of the distribution of earthquakes.

   *T/F

6. The Alps, Himalayas, and Appalachians are the result of plate collision zones.

   *T/F

7. The San Andreas Fault in California is a transform fault.

*T/F

8.The rock cycle on the continental crust is typically much faster than the rock cycle on the oceanic crust.

T/*F

9. The rock cycle and the hydrologic cycle never interact with each other.

T/*F

10. Geological rates are quite variable through time.

*T/F

11. The Mohole Project was given up for purely technical reasons.

T/*F

## COMPLETION

1. Changes in physical properties in the interior of the earth are largely controlled by **temperature** and **pressure** rather than rock composition.

2. The 3 types of plate margins are:

_____ **1. Divergent**

_____ **2. Convergent**

_____ **3. Transform Fault**

3. Molten rock is called **magma** .

4. The hydrologic cycle is powered by heat from the **sun** .

## DISCUSSION

1. Discuss the rationale for the earth being differentially layered.

2. Describe, discuss, and contrast the inner core, outer core, mesosphere, asthenosphere, and lithosphere.

3. Describe the sequence of events when a spreading center underlies a continent.

4. Describe and contrast subduction zones and collision zones.

5. Discuss the idea that rates of geological processes have varied in the past and its importance.

# CHAPTER 2

## MINERALS

### MULTIPLE CHOICE

1. The most important characteristics of minerals are _____ .

   *A - composition and crystal structure          B - crystal structure and color

   C - color and hardness          D - hardness and bonding          E - bonding and composition

2. The most fundamental substances that matter can be separated into by ordinary chemical means are _____ .

   A - minerals          *B - elements          C - bonds          D - crystals          E - B and C

3. The smallest individual particle that retains all the properties of a given chemical element.

   A - electron          B - neutron          C - proton          D - ion          *E - atom

4. The small, positively charged particles that help make atoms.

   A - electron          B - neutron          *C - proton          D - ion          E - atom

5. The small, negatively charged particles that help make atoms.

   *A - electron          B - neutron          C - proton          D - ion          E - atom

6. The small, neutral particles that help make atoms.

   A - electron          *B - neutron          C - proton          D - ion          E - atom

7. The atomic number of an element is determined by the number of _____ in the nucleus.

   A - electrons          B - neutrons          *C - protons          D - ions          E - atoms

8. An atom that has excess positive or negative charges is called a(n) _____ .

   A - electron          B - neutron          C - proton          *D - ion          E - atom

9. A cation has a _____ electrical charge.

   *A - positive          B - negative          C - neutral          D - regressive          E - progressive

10. An anion has a _____ electrical charge.

   A - positive          *B - negative          C - neutral          D - regressive          E - progressive

187

11. When electrostatic attraction draws negatively and positively charged ions together _____ bond occurs.

    *A - ionic     B - covalent   C - metallic   D - Van der Waals

    E - no such thing occurs

12. A molecule that is formed when ions share electrons has _____ bonds.

    A - ionic      *B - covalent       C - metallic   D - Van der Waals

    E - no such thing occurs

13. A special variety of covalent bonds that are closely packed ions and share electrons at inner energy-level shells are _____ bonds.

    A - ionic     B - covalent      *C - metallic  D - Van der Waals

    E - no such thing occurs

14. Compounds with _____ bonds are good conductors of electricity and heat.

    A - ionic     B - covalent      *C - metallic  D - Van der Waals

    E - no such thing occurs

15. Very weak, electrostatic bonds that do not involve transfer of electrons are _____ bonds.

    A - ionic     B - covalent      C - metallic   *D - Van der Waals

    E - no such thing occurs

16. Minerals with Van der Waals bonds tend to have a _____ feel.

    A - sticky    *B - slippery     C - harsh    D - rough      E - clammy

17. A(n) _____ ion is formed when 2 ions combine with such a strong bond that the new ion acts as a single ion.

    A - combined     B - ultra   C - fused    D - compound    *E - complex

18. In an crystalline substance the ions are _____ .

    A - random, but fixed in position      B - random, but able to move

    *C - arranged in regular, geometric patterns

19. In an amorphous substance the ions are _____ .

    *A - random, but fixed in position      B - random, but able to move

    C - arranged in regular, geometric patterns

20. In order for ionic substitution to occur the ions involved must have _____ .

    A - same electrical charge     B - distinctly different sizes  C - similar sizes

D - A and B                          *E - A and C

21. A compound that occurs in more than one crystal structure is _____ .

    A - amorphous            B - ionically substitute        *C - a polymorph

    D - mineraloid           E - non-existant

22. In 1669, Nicolaus Steno discovered the property of minerals that _____ .

    A - crystal faces are all the same size        B - size of crystal is useful in identification

    C - all minerals have a geometric relationship of hardness

    D - all minerals are water soluble

    *E - the angle between designated pairs of crystal faces is constant

23 In 1912, Max von Laue used _____ to demonstrate that crystal form reflects internal arrangement of atoms.

    A - photographs          B - the microscope           C - mass spectroscope

    *D - X-rays              E - chemical composition

24. The tendency of a mineral to break in preferred directions along bright, reflective plane surfaces is called _____ .

    A - hardness       B - luster       C - streak       *D - cleavage       E - fracture

25. The quality and intensity of light reflected from a mineral produces the effect known as _____ .

    A - hardness       *B - luster       C - streak       D - cleavage       E - fracture

26. The color produced when a thin layer of mineral powder is produced by rubbing a mineral on an unglazed porcelain plate is _____ .

    A - hardness       B - luster       *C - streak       D - cleavage       E - fracture

27. The resistance of a mineral to being scratched is _____ .

    *A - hardness       B - luster       C - streak       D - cleavage       E - fracture

28. The commonly utilized scale of hardness for minerals is named for Frederich _____ .

    A - Laus       B - Werner       *C - Mohs       D - Schmidt       E - Aaron

29. The ratio of the weight of a substance to the weight of an equal volume of pure water is _____ .

    A - density       *B - specific gravity       C - mass       D - volume weight

    E - liquid equivalent

30. Approximately _____ minerals have been described.

A - 2000      B - 2500      *C - 3000      D - 3500      E - 4000

31. Approximately _____ elements makeup 99.23% of the weight of the earth's crust.

A - 6      B - 8      C - 10      *D - 12      E - 14

32. The basic structural unit of the silicate minerals is the silica _____ .

A - diode      B - trihedron   *C - tetrahedron      D - pentahedron      E - sexagon

33. During polymerization of silica tetrahedra, 2 adjacent tetrahedra never share more than one _____ .

A - anion      B - metal ion      *C - oxygen      D - cation      E - none of the above

34. Two important mineral groups contain isolated silicate tetrahedra.

*A - olivine and garnet      B - garnet and pyroxene      C - pyroxene and amphibole

D - amphibole and micas      E - micas and olivine

35. The most important group of minerals that are single chains of silicate tetrahedra.

A - olivine      B - garnet      *C - pyroxene      D - amphibole      E - micas

36. The most important group of minerals that are double chains of silicate tetrahedra.

A - olivine      B - garnet      C - pyroxene      *D - amphibole      E - micas

36. An important group of minerals that are sheets of silicate tetrahedra.

A - olivine      B - garnet      C - pyroxene      D - amphibole      *E - micas

37. An important property of the garnets is _____ .

*A - hardness      B - luster      C - streak      D - cleavage      E - fracture

38. Among the most common minerals in the regolith are the _____ .

A - olivine      B - garnet      C - pyroxene   D - amphibole      *E - clays

39. The only common mineral composed exclusively of silicon and oxygen is _____ .

A - olivine      *B - quartz   C - pyroxene   D - amphibole      E - micas

40. The most abundant group of minerals in the earth's crust are the _____ .

A - olivine      B - quartz   C - pyroxene   *D - feldspars      E - micas

41. The feldspars are structurally are similar to quartz except some of the $Si^{4+}$ are replaced by _____ .

A - $Fe^{3+}$      B - $Cr^{3+}$      C - $Ga^{3+}$      D - $In^{3+}$      *E - $Al^{3+}$

42. The carbonate minerals have the complex _____ anion.

A - $(SO_4)^{2-}$      B - $(PO_4)^{3-}$      *C - $(CO_3)^{2-}$      D - $(NO_3)^{1-}$      E - $(CrO_4)^{2-}$

43. Two common carbonate minerals are _____ .

    A - calcite and apatite      B - apatite and gypsum      C - gypsum and anhydrite

    D - anhydrite and dolomite      *E - dolomite and calcite

43. Two common sulfate minerals are _____ .

    A - calcite and apatite      B - apatite and gypsum      *C - gypsum and anhydrite

    D - anhydrite and dolomite      E - dolomite and calcite

44. Sulfate minerals have the _____ ion.

    *A - $(SO_4)^{2-}$      B - $(PO_4)^{3-}$      C - $(CO_3)^{2-}$      D - $(NO_3)^{1-}$      E - $(CrO_4)^{2-}$

45. The most common phosphate mineral is _____ .

    A - calcite      *B - apatite      C - gypsum      D - anhydrite      E - dolomite

46. The phosphate minerals contain the _____ ion.

    A - $(SO_4)^{2-}$      *B - $(PO_4)^{3-}$      C - $(CO_3)^{2-}$      D - $(NO_3)^{1-}$      E - $(CrO_4)^{2-}$

47. Rocks can be differentiated on the basis of _____ .

    A - color and hardness      B - hardness and texture

    *C - texture and mineral assemblage

    D - mineral assemblage and shape of fragments      E - shape of fragments and color

48. The overall appearance of a rock because of the size, shape, and arrangement of its constituent mineral grains.

    A - fabric      *B - texture      C - material      D - grain      E - roughness

49. Rocks usually contain _____ .

    A - more than one kind of mineral      B - megascopic grains

    C - microscopic grains      *D - all of the above      E - only A and B

50. To study small details in a rock, _____ are often prepared.

    A - microthicknesses      B - mineral sections      C - unit slabs

    D - light transmission units      *E - thin sections

51. Igneous rocks are held by _____ .

    A - cement      *B - interlocking crystals      C - recrystallization

    D - all of the above      E - only A and C

52. Metamorphic rocks are held by _____ .

    A - cement              *B - interlocking crystals        C - recrystallization

    D - all of the above       E - only A and C

53. Sedimentary rocks may be held by _____ .

    A - cement              B - interlocking crystals        C - recrystallization

    *D - all of the above      E - only A and C

## TRUE-FALSE

1. Atoms of a chemical element are incredibly small.

    *T/F

2. All varieties of the same mineral have the same arrangement of ions.

    *T/F

3. Ionic substitution involves ions that have similar size and electrical charges.

    *T/F

4. Samples of the same mineral always look exactly alike.

    T/*F

5. The angle between cleavage planes is constant.

    *T/F

6. The mineral quartz comes in a great variety of forms.

    *T/F

7. Minerals contain the key to the condition under which they formed.

    *T/F

8. All rocks are held together with equal strength.

    T/*F

9. Virtually all minerals have an economic application.

    T/*F

10. Ore minerals are rare and hard to find.

    *T/F

## COMPLETION

1. In order for a substance to be considered a mineral it must have the following 4 characteristics.

**A--naturally formed**

**B--solid**

**C--specific chemical composition**

**D--characteristic crystal structure**

2. Rocks are aggregates of **minerals**.

3. The nucleus of an atom always contains **protons**.

4. The only common compound at the surface of the Earth that occurs in all 3 states of matter is **water** .

5. The term Mineral group is used to describe a mineral that displays extensive **ionic substitution** without changing the cation-anion ratio.

6. The physical properties of minerals are determined by **composition and crystal structure**.

7. The planar directions along which cleavage occurs are governed by the **crystal structure.**

## DISCUSSION

1. Discuss the 4 types of bonds that occur in minerals.

2. Discuss the process and importance of ionic substitution in minerals.

3. Discuss the relationship between mineral properties and type of bonding.

4. Describe how the feldspar minerals got their name.

5. Discuss the question, "What holds rock together?"

6. Discuss the question, " Can the ore minerals in the Earth's crust sustain both a growing population and a high standard of living for everyone?"

# CHAPTER 3

## THE FIRE WITHIN:
## VOLCANOS AND MAGMAS

### MULTIPLE CHOICE

1. A body of molten rock, together with any suspended mineral grains and dissolved gases, that forms when temperatures rise sufficiently high for melting to happen in the crust or mantle.

   *A - magma      B - volcano      C - lava      D - basalt      E - rhyolite

2. A vent at the earth's surface where molten rock, suspended mineral grains, and gases erupt to the surface.

   A - magma      *B - volcano      C - lava      D - basalt      E - rhyolite

3. A body of molten rock, suspended mineral grains, and gases pouring out on the earth's surface.

   A - magma      B - volcano      *C - lava      D - basalt      E - rhyolite

4. The chemical composition of magma is always predominated by _____ .

   *A - $SiO_2$      B - MgO      C - $Al_2O_3$      D - FeO      E - CaO

5. Magma is most commonly a _____ .

   A - solid      B - liquid      C - gas      D - plasma      *E - mixture of A and B

6. Magma has the general physical properties of a _____ .

   A - solid      *B - liquid      C - gas      D - plasma      E - mixture of A and B

7. A magma containing approximately 50% $SiO_2$ is _____ .

   *A - basaltic      B - andesitic      C - rhyolitic

8. A magma containing approximately 60% $SiO_2$ is _____ .

   A - basaltic      *B - andesitic      C - rhyolitic

9. A magma containing approximately 70% $SiO_2$ is _____ .

   A - basaltic      B - andesitic      *C - rhyolitic

10. Approximately 80% of all magma erupted is _____ in composition.

   *A - basaltic      B - andesitic      C - rhyolitic

11. The principal gas dissolved in magma is _____ .

A - carbon dioxide    B - nitrogen         C - chlorine    D - sulfur      *E - water vapor

12. Typically erupting volcanos are in the temperature range _____ .

    A - $600^0$ to $800^0$ C        B - $800^0$ to $1000^0$ C        *C - $1000^0$ to $1200^0$ C

    D - $1200^0$ to $1400^0$ C        E - $1400^0$ to $1600^0$ C

13. Viscosity of a magma depends on _____ .

    A - sulfur and chlorine content        B - slope and sulfur content

    C - temperature and slope

    *D - composition and temperature    E - slope and composition

14. Magmas containing 70% or more $SiO_2$ flow _____ .

    A - kilometers per hour        B - kilometers per day        C - meters per hour

    D - meters per day        *E - almost too slowly to detect

15. Magmas containing 50% or less $SiO_2$ commonly flow _____ .

    A - kilometers per hour        B - kilometers per day        *C - meters per hour

    D - centimeters per day        E - almost too slowly to detect

16. The smooth ropy-surface texture of lava that was formed from hot, gas-charged, very fluid lava is _____ .

    A - columnar joints        B - vesicular        C - aa        *D - pahoehoe

    E - massive

17. The rough-looking, rubbly surface texture of lava that was formed from cooler, gas-poor, high viscosity lava is _____ .

    A - columnar joints        B - vesicular        *C - aa        D - pahoehoe

    E - massive

18. As the concentration of $SiO_4^{4-}$ anions increases in a magma the viscosity generally increases because _____ .

    A - the anions form into irregular shaped groups of chains, sheets, etc.

    B - the anions polymerize by sharing oxygens    C - the anions too cool to be fluid

    D - all of the above                      *E - only A and B above

19. Because of the concentration of $SiO_4^{4-}$ anions, a _____ magma is always most viscous.

    *A - rhyolitic        B - andesitic        C - basaltic

20. Magma always moves toward the surface because _____ .

   A - centripetal forces of rotation throw it toward the earth's surface

   B - more gas is dissolved in the magma as it moves toward the surface

   C - the temperature of the crust increases as the surface is approached

   *D - the magma is lower-density than the containing solid rocks

   E - the pressure increases as the magma approaches the surface

21. Nonexplosive volcanic eruptions generally have magma that is _____ .

   A - high-viscosity and low dissolved-gas content

   B - high-viscosity and high dissolved-gas content

   *C - low-viscosity and low dissolved-gas content

   D - low-viscosity and high dissolved-gas content

   E - none of the above

22. When bits of falling lava pile up around a vent a _____ forms.

   A - ramparted surface          B - pahoehoe slope          *C - spatter cone

   D - all of the above           E - only A and C

23. Bubble holes preserved in an solidified lava flow are called _____ .

   A - amygdules      B - vugs      C - trails      D - capillaries      *E - vesicles

24. Bubble holes that are filled by secondary minerals in an solidified lava flow are called _____ .

   *A - amygdules      B - vugs      C - trails      D - capillaries      E - vesicles

25. When escaping gases form a mass of bubbles in a rhyolitic magma the rock _____ forms.

   A - vesicular rhyolite          *B - pumice          C - obsidian

   D - amygdaloidal rhyolite       E - pitchstone

26. Rocks formed from fragments ejected from an erupting volcano are _____ .

   *A - pyroclastic      B - tephra      C - pumice      D - obsidian      E - basalt

27. Volcanic bombs, lapilli, and ash differ in _____ .

   A - density      B - composition      *C - size      D - color      E - age

28. Eruption columns form when _____ .

   A - magma cools in the mouth of an erupting volcano

196

B - lava flow cools and shrinks     C - ignimbrites cool

D - a hot, highly mobile flow of tephra rushes down the flank of a volcano

*E- mixtures of hot, turbulent gas and rock fragments erupt into the air

29. Pyroclastic flows occur when _____ .

   A - magma cools in the mouth of an erupting volcano

   B - lava flow cools and shrinks     E - ignimbrites cool

   *C - a hot, highly mobile flow of tephra rushes down the flank of a volcano

   D - mixtures of hot, turbulent gas and rock fragments erupt into the air

30. Rock formed from a poorly sorted mass of blocks, lapilli, and ash is _____.

   *A - ignimbrite     B - eruption column tuff     C - pumice     D - obsidian

   E - basalt

31. The _____ is the most easily visualized type of volcano. It is one built up of numerous thin flows of nearly uniform thickness.

   A - cinder cone     B - tephra cone     C - stratovolcano     *D - shield volcano

   E - lava dome

32. Hawaii, Tahiti, Samoa, and the Galapagos islands are examples of _____ .

   A - cinder cone     B - tephra cone     C - stratovolcano     *D - shield volcano

   E - lava dome

33. A rhyolitic and andesitic volcano that produce large amounts of pyroclasts produce steep sided a _____ .

   A - spatter cone     *B - tephra cone     C - stratovolcano     D - shield volcano

   E - lava dome

34. Large, long-lived volcanos that emit interbedded lava flows and tephra falls are _____ .

   A - spatter cone     B - tephra cone     *C - stratovolcano     D - shield volcano

   E - lava dome

35. Stratovolcanos are differentiated from tephra cones because they have _____ .

   A - pyroclastic     B - tephra     C - pumice     D - obsidian     *E - lava flows

36. The principal gas emitted by a volcano is _____ .

   A - carbon dioxide     B - nitrogen     C - chlorine     D - sulfur     *E - water vapor

37. Old magma chambers tend to remain hot for _____ of years.

A - hundreds          B - thousands          C - tens of thousands

*D - hundreds of thousands          E - millions

38. A thermal spring equipped with a system of plumbing and heating that causes intermittent eruptions of water and steam is a _____ .

A - artesian   *B - geyser   C - fissure   D - spatter spring   E - tephra spring

39. A "small" funnel-shaped depression located near the summit of a volcano from which materials are ejected is a _____ .

A - caldera   B - geyser   *C - crater   D - magma subduction depression

E - tephra ejection structure

40. A "large", roughly circular, steep-walled basin several kilometers in diameter located near the summit of a volcano from which materials are ejected is a _____ .

*A - caldera   B - geyser   C - crater   D - magma subduction depression

E - tephra ejection structure

41. When magma starts to reenter the chamber under a caldera and causes the floor of the caldera to rise, the structural dome formed is called a _____ .

A - lava dome          B- tephra dome          *C - resurgent dome

D - plateau dome          E - pillow dome

42. When sticky, viscous lava with little dissolved gas is extruded following a major volcanic eruption it is called a _____ .

*A - lava dome          B- tephra dome          C - resurgent dome

D - plateau dome          E - pillow dome

43. Plateau basalts are the result of _____ .

A - rapid uplift of the area of eruption          B - stratovolcanos          C - lava domes

D - pillow eruptions          *E - fissure eruptions

44. The most extensive volcanic system on the Earth occurs along _____ .

A - the crest of the Rocky Mountains          B - oceanic trenches   *C - midocean ridges

D - along continental margins          E - the east side of the Andes Mountains

45. Both extrusive and intrusive igneous rocks are classified and names on the basis of _____ .

A - color and rock texture          B - density and color          C - age and density

D - mineral assemblage and age          * E - rock texture and mineral assemblage

46. Extrusive igneous rocks are formed by the process of _____ .

A - magma solidifying within the crust or mantle

B - accumulation of fragments at the surface of the earth

\* C - cooling and solidification of lava

D - nondestructive alteration of crustal material

E - alteration of crustal materials in a solid state

47. Intrusive igneous rocks are formed by the process of _____ .

\*A - magma solidifying within the crust or mantle

B - accumulation of fragments at the surface of the earth

C - cooling and solidification of lava

D - nondestructive alteration of crustal material

E - alteration of crustal materials in a solid state

48. Intrusive igneous rocks generally are _____ grained.

    A - fine    \*B - coarse    C - glassy    D - extremely coarse    E - A or C

49. Extrusive igneous rocks generally are _____ grained.

    \*A - fine    B - coarse    C - glassy    D - extremely coarse    E - A or C

50. A texture with coarse grains scattered through a finer grained groundmass is a _____ .

    A - phenocryst        B - pegmatite        C - mixogranular

    \*D - porphyry        E - granulite

51. A two stage cooling history in an igneous rock is indicated by a _____ texture.

    A - phenocryst        B - pegmatite        C - mixogranular

    \*D - porphyry        E - granulite

52. The glassy texture of obsidian indicates _____ so atoms do not have time to organize themselves into minerals.

    A - cooling under extreme pressure        B - cooling extremely slowly

    C - cooling in the presence of a mineraloid    \*D - cooling extremely rapidly

    E - cooling in the absence of $SiO_4^{4-}$ anions

53. An intrusive igneous rock that contains unusually large (greater than 2 cm) mineral grains.

    A - phenocryst        \*B - pegmatite        C - mixogranular

    D - porphyry        E - granulite

54. The common igneous light colored minerals are _____.

    A - quartz, olivine, muscovite          B - biotite, amphibole, pyroxene, olivine

    C - biotite, quartz, pyroxene          *D - quartz, feldspar, muscovite

    E - quartz, biotite, olivine

55. The common igneous dark colored minerals are _____.

    A - quartz, olivine, muscovite          *B - biotite, amphibole, pyroxene, olivine

    C - biotite, quartz, pyroxene          D - quartz, feldspar, muscovite

    E - quartz, biotite, olivine

56. An igneous rock that is dominantly quartz and feldspar.

    *A - granite    B - diorite    C - gabbro    D - peridotite      E - amphibolite

57. An igneous rock that is dominantly plagioclase, but no quartz.

    A - granite    *B - diorite    C - gabbro    D - peridotite      E - amphibolite

58. An igneous rock that is more than 50% pyroxene and olivine.

    A - granite    B - diorite    *C - gabbro    D - peridotite      E - amphibolite

59. An igneous rock that 90% or more olivine.

    A - granite    B - diorite    C - gabbro    *D - peridotite      E - amphibolite

60. The dominant rock on the ocean floors is _____ .

    A - rhyolite    B - dacite    C - andesite      *D - basalt    E - none of the above

61. All bodies of intrusive igneous rocks are called _____ regardless of origin.

    A - necktons   B - benthons      C - planktons      *D - plutons   E - vulcans

62. An intrusive igneous rock body that is tabular in form, parallel-sided sheet that cuts across the layering of the rock it intrudes.

    A - sill      *B - dike      C - stock      D - batholith      E - laccolith

63. An intrusive igneous rock body that is tabular in form, parallel-sided sheet that is parallel to the layering of the rock it intrudes.

    *A - sill      B - dike      C - stock      D - batholith      E - laccolith

64. An igneous body intruded parallel to the layering of the rock it intrudes and above which the layers of the intruded rocks have been bent upward to form a dome.

    A - sill      B - dike      C - stock      D - batholith      *E - laccolith

65. The largest pluton, irregular in shape that cuts across the layering it intrudes.

    A - sill      B - dike      C - stock      *D - batholith      E - laccolith

66. An irregularly shaped pluton that cuts across the layering it intrudes, but is less than 10 km maximum dimension.

   A - sill      B - dike      *C - stock      D - batholith      E - laccolith

67. A block of host rock that has been stoped from the walls during emplacement of an intrusive rock body, but not dissolved by the magma.

   A - autolith   B - stoplith   *C - xenolith      D - gravolith      E - stocklith

68. Rhyolitic volcanos most abundantly occur _____ .

   A - at spreading centers of the lithospheric plates

   B - landward, but parallel to the zones of subduction of lithospheric plates

   *C - on continental crust      D - at intraplate hot spots      E - both A and C

69. Andesitic volcanos most abundantly occur _____ .

   A - at spreading centers of the lithospheric plates

   *B - landward, but parallel to the zones of subduction of lithospheric plates

   C - on continental crust      D - at intraplate hot spots      E - both A and C

70. Basaltic volcanos most abundantly occur _____ .

   A - at spreading centers of the lithospheric plates

   B - landward, but parallel to the zones of subduction of lithospheric plates

   C - on continental crust      D - at intraplate hot spots      *E - both A and C

71. The Andesite Line is _____ .

   A - exactly parallel to the plate subduction margins

   B - exactly parallel to the plate spreading margins

   C - around the Pacific Ocean            *D - A and C            E - B and C

72. Almost 200 years ago, James Hutton demonstrated that _____ .

   A - the geothermal gradient is the same under continental and oceanic crust

   *B - rocks can be melted      C - volcanos occur only on oceanic crust

   D - the Hawaiian Islands lie over a hot spot

   E - Iceland lies on the midoceanic ridge

73. The geothermal gradient is the rate of change of temperature _____ .

   A - as you proceed poleward from the equator      B - as you cross a mountain range

   C - as a rock melts at the Earth's surface      *D - as you deeper and deeper into the earth

E - as lava cools

74. The melting point of a mineral lowers as _____ .

    A - pressure increases and water(vapor) decreases

    B - pressure decreases and water(vapor) decreases

    *C - pressure decreases and water(vapor) increases

    D - pressure increases and water(vapor) increases        E - none of the above

75. Partial melting of rocks occur because _____ .

    A - there is too much water present        B - there is too little water present

    C - there is not enough time for complete melting        D - the pressure is too low

    *E - rocks are a mixture of minerals

76. Basaltic magma appears to form _____ .

    A - under extremely dry melting conditions

    B - from a 10% to 15% partial melting of the garnet peridotite mantle

    C - temperature-pressure conditions similar to those in the Asthenosphere

    *D - all of the above        E - only A and B

77. Andesitic magma appears to form _____ .

    A - under wet conditions        B - because of partial melting of oceanic crust

    C - in the trenches        D - all of the above        *E - only A and B

78. Rhyolitic magma appears to form _____ .

    A - under wet conditions        B - because of partial melting of continental crust

    C - in the crust        *D - all of the above        E - only A and B

79. Rhyolitic magma _____ .

    *A - most commonly cools within the crust to form granitic batholiths

    B - usually erupts to form shield volcanos

    C - stay hot for long periods of time allowing it to easily traverse the crust

    D - is extremely dry initially

    E - represents the only complete melting sequence in the crust

80. Magmatic differentiation by fractional crystallization occurs because _____ .

    A - mineral melt at different temperatures

B - the amount of water varies from magma to magma

*C - minerals crystallize at different temperatures

D - magmas vary in composition          E - there is no such process

81. Magmatic differentiation by fractional crystallization occurs because _____ .

A - compression can squeeze melt out of a crystal-melt mixture

B - gravity causes early formed minerals to settle to the bottom

C - buoyancy causes early formed mineral to float to the top

*D - A and B                    E - A and C

82. Bowen's Reaction Series has _____ .

A - a continuous reaction series of olivine, pyroxene, amphibole, biotite

B - a discontinuous reaction series of olivine, pyroxene, amphibole, biotite

C - a continuous reaction series of plagioclase feldspars

D - a discontinuous reaction series of plagioclase feldspars          *E - A and C

83. Magmatic mineral deposits may form _____ .

A - very late in crystallization     B - from residual magma with concentrated elements

C - because of early crystal settling       *D - all of the above    E - only A and C

## TRUE-FALSE

1. In general, the higher the temperature, the lower the viscosity and the more readily a magma flows.

*T/F

2. As the content of $SiO_4^{4-}$ anions increases in a magma the viscosity generally decreases.

T/*F

3. At higher pressures, a magma will hold less dissolved gas than at lower pressures.

T/*F

4. Spatter cones, pahoehoe, and aa may all form from the same Hawaiian-type volcanic eruption.

*T/F

5. The rock pumice floats because it has so many holes in it.

*T/F

6. Geologists were prepared when Mt. St. Helens erupted in 1980.

   T/*F

7. Mount Baker, Mount Rainier, and Mount Hood are excellent examples of tephra cones.

   T/*F

8. Mount Baker, Mount Rainier, and Mount Hood are excellent examples of stratovolcanos.

   *T/F

9. Most of the world's geysers outside of Iceland are in New Zealand and Yellowstone National Park.

   *T/F

10. Most of the world's geysers outside of Iceland are in New York and Grand Canyon National Park.

    T/*F

11. Volcanic craters are much larger that volcanic calderas.

    T/*F

12. The largest fissure eruption in historic times was at Laki, Iceland.

    *T/F

13. The largest fissure eruption in historic times was the Roza Flow in Washington State.

    T/*F

14. Most of the lavas on the oceanic crust are pillow basalts.

    *T/F

15. The boundaries between the several igneous rocks are quite distinct.

    T/*F

16. The Hawaiian Islands seem to lie over a hot spot.

    *T/F

17. The geothermal gradient in and beneath the continental crust is the same as the geothermal gradient in and beneath the oceanic crust.

    T/*F

18. All minerals melt at the same temperature.

    T/*F

19. The chemical composition of andesite magma is close to the average composition of the continental crust.

    *T/F

20. Rhyolitic magma rises rapidly because it is extremely fluid.

    T/*F

21. A magma of a given composition can crystallize into many different kinds of igneous rocks.

    *T/F

22. Geothermal energy is relatively unusual and difficult to find.

    *T/F

# DISCUSSION

1. List and discuss 5 volcanic hazards discussed in the text.

2. discuss the procedure for naming igneous rocks.

3. Discuss the conversion of tephra to pyroclastic rock.

4. Discuss and describe the distribution of rhyolitic, andesitic, and basaltic volcanos and place of origin of the magma for each.

5. Why isn't the Earth's mantle entirely molten?

# CHAPTER 4

## SEDIMENTS AND SEDIMENTARY ROCKS

### MULTIPLE CHOICE

1. Sedimentary rocks have _____ .

    A - stratification          B - bedding          C - bedding planes

    *D - all of the above       E - only A and B

2. The top or bottom surface of a layer of rock _____ .

    A - stratification          B - bedding          *C - bedding plane

    D - all of the above        E - only A and B

3. Pebbles and sand grains are usually _____ .

    A - rock fragments          B - chemical precipitates          C - mineral grains

    D - all of the above        *E - only A and C

4. Clastic sediments are classified mainly on the basis of _____ .

    A - color       B - shape       C - composition       *D - size clasts

    E - type of cement

5. Clastic particles are produced by _____ .

    A - erosion     *B - weathering       C - mass wasting       D - all of the above

    E - none of the above

6. Clasts may be composed of _____ .

    A - fragments of igneous rock       B - mineral grains     C - fragments of shells

    D - fragments of metamorphic rock        *E - all of the above

7. Deposition occur because of _____ of the transporting agent.

    A - an increase in energy          B - wear on the clasts          C - a change of phase

    *D - a decrease in energy          E - only A and C

8. The most resistant of the common rock forming minerals to weathering is _____ .

    A -  potassium feldspar          B - plagioclase          *C - quartz

D - mica                              E - amphibole

9. In general, the greater the rate of speed of the transporting medium, the _____ size of the clast being transported.

   *A - greater                 B - smaller            C - makes no difference

10. Energy fluctuations in the transporting medium of clasts are _____ .

    A - large and unusual        B - small and unusual        *C - small and common

    D - large and common                E - nonexistent

11. Clastic particles are typically sorted according to _____ .

    A - specific gravity    *B - size        C - color        D - shape        E - mineralogy

12. Mechanically weathered particles broken from bedrock tend to be angular because breakage typically occurs along _____ .

    A - grain boundaries        B - fractures            C - surfaces separating rock layers

    *D - all of the above        E - only A and B

13. The measure of the sharpness of a particle's edges is _____.

    *A - roundness              B - sphericity            C - smoothness

    D - wearing                 E - hardness

14. An annually accumulated couplet of sediment, typically associated with glacial sedimentation.

    A - cross bedding      B - graded bedding      C - till        *D - varve

    E - all of the above

15. Beds that are inclined with respect to a thicker stratum with respect to a thicker stratum within which they occur.

    *A - cross bedding      B - graded bedding      C - till        D - varve

    E - all of the above

16. Particles sorted by size and grading upward upward from coarser to finer.

    A - cross bedding      *B - graded bedding      C - till        D - varve

    E - all of the above

17. Chemical sediment may form as a result of _____ .

    A - biogenic activity        B - inorganic activity

    C - accumulation of unimineral clasts        D - all of the above      *E - only A and B

18. The remains of plants and animals that died and were incorporated and preserved as the sediment accumulated.

A - biogens     B - pathogens          C - bioclasts     *D - fossils     E - none of the above

19. Biogenic sediment is dominated by _____.

A - calcareous          B - siliceous          C - phosphatic          *D - A and B

E - B and C

20. During the process of photosynthesis, plants utilize _____ in the production of carbohydrates and oxygen.

A - water     B - ammonia     C - carbon dioxide     D - all of the above

*E - only A and C

21. The first stage in the development of solid fossil fuels is _____ .

*A - peat          B - lignite          C - coal          D - anthracite          E - cannel

22. The most abundant fossil fuel is _____.

A - peat          B - lignite          *C - coal          D - anthracite          E - cannel

23. Petroleum appears to be derived from _____ .

A - marine microphytoplankton          B - marine bacteria

C - marine microzooplankton          D - all of the above          *E - only A and B

24. Petroleum may occur in _____ form.

A - gaseous     B - liquid     C - semisolid          *D - all of the above     E - A and B

25. The collective term used to describe all of the chemical, physical, and biological changes that affect sediment after its initial deposition, and during and after lithification.

*A - diagenesis          B - solid alteration          C - lithogenesis

D - petrafaction          E - regeneration

26. The process whereby a newly deposited, unconsolidated sediment is slowly converted to sedimentary rocks.

A - diagenesis          B - solid alteration          *C - lithification

D - petrafaction          E - regeneration

27. Sedimentary rocks made from igneous rock fragments can be distinguished from igneous rocks by _____ .

A - rounded fragments cemented together          B - presence of fossils

C - interlocking grains          D - all of the above          *E - only A and B

28. Clastic sedimentary rock composed of the smallest size fragments.

A - conglomerate    B - sandstone    C - siltstone    *D - shale    E - greywacke

29. Clastic sedimentary rock composed predominantly of quartz fragments.

A - conglomerate    *B - sandstone    C - siltstone    D - shale    E - greywacke

30. Clastic sedimentary rock composed predominantly of lithic fragments.

A - conglomerate    B - sandstone    C - siltstone    D - shale    *E - greywacke

31. Most of the world's iron ore is mined from _____ .

A - igneous veins    B - magmatic alteration zones    *C - bedded iron deposits

D - placer iron    E - metamorphic halos

32. Among the common evaporites that have economic value are _____ .

A - quartz, mica, calcite    B - amphibole, olivine, plagioclase

C - hematite, limonite, goethite    D - almandite, staurolite, kyanite

*E - gypsum, halite, carnallite

33. A major portion of the $CO_2$ stored in the earth's crust is contained in the rock _____ .

A - shale    B - sandstone    C - rock salt    *D - limestone    E - slate

34. Coarse grained, poorly cemented limestone composed of shelly material.

*A - coquina    B - chalk    C - oölitic limestone

D - lithographic limestone    E - reef limestone

35. Limestone composed of cemented reef material.

A - coquina    B - chalk    C - oölitic limestone

D - lithographic limestone    *E - reef limestone

36. Limestone composed of small, spherical, accretionary bodies.

A - coquina    B - chalk    *C - oölitic limestone

D - lithographic limestone    E - reef limestone

37. Limestone composed of chemically precipitated lime mud.

A - coquina    B - chalk    C - oölitic limestone

*D - lithographic limestone    E - reef limestone

38. Limestone composed of shells of minute floating organisms.

A - coquina    *B - chalk    C - oölitic limestone

D - lithographic limestone    E - reef limestone

39. Diatomite is used for _____ .

    A - filters     B - abrasive        C - paint filler      D - all of the above

    *E - only A and B

40. Oil shales have a waxlike substance instead of petroleum because _____ .

    A - no solvents were present                B - the rocks are too young

    *C - the rocks were not heated hot enough      D - the rocks are too fine grained

    E - all of the above

41. Dark colored sedimentary rocks usually contain _____ as a coloring agent.

    A - iron sulfides     B - amphibole       C - organic matter    D -A and B

    *E - A and C

42. Reddish and brownish colors in rocks is usually due to _____ .

    A - garnets   *B - iron oxide     C - organic debris    D - copper

    E - manganese silicate

43. Stream deposited sediment differs from place to place depending on _____ .

    A - type of stream       B - the amount of energy available to do work

    C - nature of the sediment load       *D - all of the above    E - only A and B

44. Sediment directly deposited by glaciers is typically _____ .

    A - very simple in composition        B - unsorted        C - unstratified

    D - all of the above           *E - only B and C

45. Wind erosion most readily occurs where _____ .

    A - strong wind occurs       B - vegetation is discontinuous

    C - small sized sediment is present     *D - all of the above    E - only A and B

46. Under normal depositional conditions sediment gets _____ going away from the source.

    A - coarser          B - uniformly mixed         *C - finer

    D - more siliceous       E - more organic matter

47. Beach gravel tends to acquire a more _____ shape than similar gravelly alluvium.

    A - spherical      B - rounded   C - angular   D - cylindrical     *E - flattened

48. Most coarse grained marine sediment is deposited within _____ km of the land after being dispersed by currents.

A - 1-2          B - 3-4          *C - 5-6          D - 7-8          E - 9-10

49. During the recent geologic past, sea level has changed several times. As much as
_____% of the sediment on the continental shelves is relict from this.

A - 40          B - 50          C - 60          *D - 70          E - 80

50. Approximately _____% of the sediment that reached the continental shelves is
deposited there.

A - 50          B - 60          C - 70          D - 80          *E - 90

51. Biogenic carbonate sediments tend to accumulate where _____ .

    A - influx of land sediments is minimal          B - climate is warm

    C - sea-surface temperature is warm          *D - all of the above    E - none of the above

52. Turbidity currents are _____ .

    A - gravity driven          B - dilute mixtures of sediment and water

    C - highly regular          D - all of the above          *E - only A and B

53. Turbidity currents may be set off by _____ .

    A - earthquakes          B - landslides          C - major coastal storms

    *D - all of the above          E - only A and B

54. Sediment for deep-sea fans is derived mostly from _____ .

    *A - the lands          B - carbonate banks          C - abyssal plains

    D - all of the above          E - only A and B

55. Siliceous ooze is most common is _____ .

    A - equatorial Atlantic Ocean          B - equatorial Pacific Ocean

    C - equatorial Indian Ocean          D - all of the above

    *E - only B and C

56. Calcareous ooze does not occur at depths greater than approximately _____ km in
ward latitudes.

A - 2          B - 3          *C - 4          D - 5          E - 6

57. In the North Pacific and parts of the central South Pacific where biogenic ooze does not
accumulate, the ocean floor is covered by _____ .

    A - barren black and dark green basalt          *B - oxidized reddish or brownish clay

    C - green authegenic minerals          D - all of the above          E - only A and C

58. Energy to drive sedimentation comes ultimately from _____ .

    A - internal heat of the Earth       B - the Sun       C - atmospheric pressure

    D - all of the above       *E - only A and B

59. In tectonically active regions, rates of uplift are _____ rates of sedimentation.

    *A - greater than       B - less than       C - same as

60. Along the splitting edge of a continent, great thicknesses of _____ accumulate.

    *A -shallow water sediments       B - coarse stream sediments

    C - volcanic rich sediments       D - calcareous sediments

    E - oxidized reddish or brownish clay

61. Along the colliding edge of a continent, great thicknesses of _____ accumulate.

    A -shallow water sediments       *B - coarse stream sediments

    C - volcanic rich sediments       D - calcareous sediments

    E - oxidized reddish or brownish clay

62. Along the subducting edge of a continent, great thicknesses of _____ accumulate.

    A -shallow water sediments       B - coarse stream sediments

    *C - volcanic rich sediments       D - calcareous sediments

    E - oxidized reddish or brownish clay

## TRUE-FALSE

1. The ultimate resting place of most of the sediment eroded from the land is the world's oceans.

    *T/F

2. During transportation, detrital grains are not subjected to further alteration.

    T/*F

3. The preserved remains of plants and animals are usually found complete and intact in sediments.

    T/*F

4. Most carbonate sediments result from biogenic activity.

    *T/F

5. All oils chemically alike.

    T/*F

6. Deposition of phosphorus appears to have been much more common in the geologic past.

    *T/F

7. All stream deposited sediments are alike.

    T/*F

8. Typically wind deposited dunes are easily identified in the rock record.

    *T/F

9. Wind deposited silt is common in the rock record.

    T/*F

10. When fresh water mixes with sea water, the clays tend to clump together and settle faster.

    *T/F

11. Sedimentation in a large delta is very simple.

    T/*F

12. Ocean beaches are all sandy.

    T/*F

13. Sea level has been constant throughout the history of the earth.

    T*/F

14. Marine evaporites are relatively widespread in the rock record.

    *T/F

15. Turbidity currents happen often at any given site.

    T/*F

16. Deep-sea sediments typically have a single source.

    T/*F

## COMPLETION

1.The loose, fragmental debris produced by the mechanical breakdown of older rocks is **detritus**.

2. The chief mechanism for trapping solar energy on earth is **photosynthesis**.

3. List the 3 common diagenetic processes.

**A--compaction**

**B--cementation**

**C--recrystallization**

4. Rock salt and gypsum probably originate by **evaporation** of sea water.

5. The body of sediment deposited at the mouth of a stream where it enters a lake or ocean is a **delta**.

# DISCUSSION

1. Discuss the character and origin of varves.

2. Discuss the origin of chemical sediments.

3. Discuss 2 lines of evidence that support the hypothesis that petroleum is a product of the decomposition of organic matter.

4. Discuss the evidence that coal plants were swamp dweller and the order of events and conditions that lead from peat to coal.

5. Discuss the chemical conditions controlling the origin of bedded iron deposits.

6. Discuss the types of environmental data that can be derived from the study of fossils.

7. Discuss the relationship between tectonics and sedimentation.

# CHAPTER 5

## NEW ROCKS FROM OLD:
## METAMORPHISM AND METAMORPHIC ROCKS

### MULTIPLE CHOICE

1. During metamorphism, the rocks are in a _____ state.

    A - liquid                          B - gas                       *C - solid

2. Minimum conditions for metamorphism to begin are _____ .

    A - 150$^0$ C and 200 Mpa    B - 200$^0$ C and 200 Mpa    *C - 200$^0$ C and 300 Mpa

    D - 250$^0$ C and 300 Mpa    E - 300$^0$ C and 350 Mpa

3. The upper limit of metamorphism is when _____ melting occurs.

    A - dry complete           B - dry partial           C - wet complete

    *D - wet partial           E - none of the above

4. As the amount of water present in rocks increases, the melting point _____ .

    A - increases           *B - decreases           C - is uneffected

5. Igneous and metamorphic rocks are normally found in close association. The plate tectonic setting for the association is _____ plate boundary.

    A - diverging        B - collision        C - subduction        D - all of the above

    *E - only B and C

6. The final metamorphic rock produced is largely controlled by _____ .

    A - initial composition of the starting rock        B - temperature of the process

    C - pressure of the process        *D - all of the above        E - only A and B

7. Intergranular fluids are water with small amounts of _____ .

    A - dissolved gases    B - salts    C - mineral constituents of enclosing rocks

    *D - all of the above        E - only A and B

8. The presence of intergranular fluids _____ the rate of metamorphic reactions.

    *A - speeds-up          B - slows-down          C - do not effect

9. As depth of burial increases, the amount of pore space _____ .

A - increases          *B - decreases          C - remains constant

10. Prograde metamorphic effects occur while _____ .

   A - pressure and temperature are rising     B - pressure and temperature are falling

   C - intergranular pore fluids are abundant   *D - only A and C     E - only B and C

11. Retrograde metamorphic effects occur while _____ .

   A - pressure and temperature are rising     B - pressure and temperature are falling

   C - intergranular pore fluids have been expelled      D - only A and C

   *E - only B and C

12. Metamorphic rocks typically reflect _____ stress with _____ crystals.

   A - homogeneous/oriented          B - homogeneous/randomly

   *C - differential/oriented     D - differential/randomly     E - none of the above

13. Igneous rocks typically reflect _____ stress with _____ crystals.

   A - homogeneous/oriented          *B - homogeneous/randomly

   C - differential/oriented     D - differential/randomly     E - none of the above

14. When two chemical compounds combine, the resultant compound must be at a _____ .

   A - higher energy state than either starting compound

   *B - lower energy state than either starting compound

   C - intermediate energy state to either starting compound

   D - indeterminate energy state to either starting compound

   E - unpredictable energy state to either starting compound

15. Metamorphic rocks that tend to split into thin, leaflike flakes are described as _____ .

   A - striated     B - layered     C - leafed     *D - foliated          E - texturized

16. Most metamorphic rocks form in a _____ stress field and as a result they develop
conspicuous, directional textures.

   *A - nonhydrostatic          B - hydrostatic          C - nonpneumatic

   D - pneumatic          E - none of the above

17. During the very earliest stages of low-grade metamorphism, stress tends to be caused by
_____ .

   A - heat of burial          *B - the weight of the overlying rock

   C - subduction processes     D - nearby magmas          E - plate collision processes

18. The property by which a low-grade metamorphic rock breaks into platelike fragments along planes parallel to microscopic crystals is _____ .

    A - mineral cleavage             B - schistosity            *C - slaty cleavage

    D - differential breakage        E - platelet repetition

19. Schistosity differs from slaty cleavage mainly in _____ .

    A - the minerals involved         B - the direction of cleaving

    C - thickness of cleaving units     *D - grain size        E - strength of bonds

20. What does the presence of chlorite, serpentine, epidote, and talc indicate about the host rock?

    A - it is redeposited igneous       B - it is an arid climate sedimentary

    C - it is normal igneous           D - it has undergone intense weathering

    *E - it is metamorphic

21. Kyanite, sillimanite, and andalusite are _____ .

    *A - polymorphs     B - typical sedimentary rocks      C - common in igneous rocks

    D - unrelated to each other      E - never found together

22. For a given chemical composition, any given temperature-pressure range will _____ .

    A - change the schistosity    B - have no effect

    C - start a different reaction sequence    *D - produce a specific mineral assemblage

    E - change the orientation of the grains

23. The naming of metamorphic rocks is based partly on_____, and partly on_____.

    A - color/mineral assemblage        B - texture/orientation of grains

    *C - texture/mineral assemblage       D - texture/specific gravity

    E - specific gravity/mineral assemblage

24. The low-grade metamorphic product of either shale or mudstone.

    *A - slate      B - phyllite    C - schist     D - gneiss     E - marble

25. The presence of _____ is clear proof that a rock has gone from being a sedimentary rock to a metamorphic rock.

    A - a color change       B - higher specific gravity     *C - slaty cleavage

    D - all of the above       E - only A and C

26. Intermediate grade metamorphism with pronounced foliation and grains just large enough to be visible.

A - slate      *B - phyllite    C - schist     D - gneiss     E - marble

27. Coarse grained metamorphic rock with pronounced schistosity.

     A - slate      B - phyllite    *C - schist     D - gneiss     E - marble

28. A high-grade metamorphic rock with coarse grains and pronounced foliation, but with layers of micaceous minerals segregated from the other minerals.

     A - slate      B - phyllite    C - schist     *D - gneiss     E - marble

29. The most obvious differences between slate, phyllite, and schist are _____ .

     A - color      B - basic chemistry    C - foliation    *D - grain size

     E - parent rock

30. When basalt metamorphoses in the presence of water, the lowest grade rock formed is _____ .

     *A - greenschist        B - amphobolite        C - pyroxeneite

     D - serpentineite        E - granulite

31. The color in greenschist comes from the presence of ____ .

     A - plagioclase      B - epidote    C - calcite     D - pyroxene      *E - chlorite

32. The principle mineral in granulite is _____ .

     A - plagioclase      B - epidote    C - calcite     *D - pyroxene     E - chlorite

33. As basalt is metamorphosed to higher grade this is the order of dominant mineral in the resulting rock type.

     A - pyroxene - chlorite - amphobole      *B - chlorite - amphibole - pyroxene

     C - chlorite - pyroxene - amphibole      D - pyroxene - amphibole - chlorite

     E - amphibole - chlorite - pyroxene

34. Marble and quartzite commonly lack _____

     A - bedding planes      B - foliation      C - original grain structure

     *D - all of the above      E - none of the above

35. Metamorphic rock consisting of coarsely crystalline, interlocking network of calcite grains.

     A - slate      B - phyllite    C - schist     D - quartzite      *E - marble

36. The parent rock for marble is _____ .

     A - shale     B - sandstone      C - basalt      *D - limestone

E - conglomerate

37. The parent rock for quartzite is _____ .

   A - shale      *B - sandstone      C - basalt      D - limestone

   E - conglomerate

38. The parent rock for greenschist is _____ .

   A - shale      B - sandstone      *C - basalt      D - limestone

   E - conglomerate

39. The parent rock for schist is _____ .

   *A - shale      B - sandstone      C - basalt      D - limestone

   E - conglomerate

40. Mechanical deformation without any accompanying chemical recrystallization.

   *A - cataclastic metamorphism          B - regional metamorphism

   C - burial metamorphism               D - contact metamorphism

41. Chemical recrystallization adjacent to bodies of hot magma.

   A - cataclastic metamorphism          *B - contact metamorphism

   C - regional metamorphism             D - burial metamorphism

42. A well defined shell of altered rock around an intrusion.

   A - metamorphic vacuole    *B - metamorphic aureole    C - metamorphic halo

   D - metamorphic glow       E - metamorphic capstone

43. A hard, fine-grained, baked metamorphic rock composed of an interlocking mass of uniformly sized mineral grains.

   A - slate      B - phyllite      C - schist      D - quartzite    *E - hornfels

44. Metamorphic rock that looks like the unaltered sedimentary rock and retains the texture, but has a completely different mineralogy.

   A - contact metamorphism          B - cataclastic metamorphism

   *C - burial metamorphism          D - regional metamorphism

45. As temperatures and pressures increase burial metamorphism grades into _____ .

   *A - regional metamorphism         B - burial metamorphism

   C - contact metamorphism           D - cataclastic metamorphism

46. Metamorphic rocks that are distinctly foliated as the result of differential stress, a considerable amount of mechanical deformation plus chemical recrystallization.

*A - regional metamorphism      B - burial metamorphism

C - contact metamorphism      D - cataclastic metamorphism

47. A line on a map connecting points of first occurrence of a given mineral in metamorphic rocks.

A - isobar      B - isolith      C - isomin      *D - isograd

E - all of these words are imaginary

48. The first systematic study of a regionally metamorphosed terrain was carried-out in _____ .

A - Finland      *B - Scottish Highlands      C - Canadian Shield

D - western Russia      E - the Adirondack Mountains

49. The regions on a map between isograds are _____ .

*A - metamorphic zones      B - metamorphic areas      C - index regions

D - index mineral areas      E - area zones

50. The main changes that occur to rocks during metamorphism is the loss or addition of _____ .

A - $H_2O$ and $SO_3$      B - $H_2O$ and $NO_2$      *C - $H_2O$ and $CO_2$

D - $H_2O$ and $NH_3$      E - $H_2O$ and $CH_4$

51. The concept that for a given rock composition, the assemblage of minerals that reach equilibrium during metamorphism within a specific range of physical conditions belong to the same _____ .

A - zone region      B - facies zone      C - isograd zone

*D - metamorphic facies      E - metamorphic region

52. When Eskola originally described the concept of metamorphic facies he was studying metamorphosed _____ .

A - shale      B - sandstone      *C - basalt      D - limestone  E - conglomerate

53. Most metamorphic processes involve relatively small amounts of water.  The typical water:rock ratio is approximately _____ or less.

A - 1:25      B - 1:20      C - 1:15      *D - 1:10      E - 1:5

54. The process whereby rocks have their chemical composition distinctively altered by the addition or removal of ions in solution.

A - hydromorphism      *B - metasomatism      C - metahydrolysis

D - somatichydrolysis      E - morphosomatism

55. Metasomatism is commonly associated with _____ .

    A - cataclastic metamorphism and contact metamorphism

    *B - contact metamorphism and regional metamorphism

    C - regional metamorphism and burial metamorphism

    D - burial metamorphism and cataclastic metamorphism

56. Most fluid that cause metasomatism tend to be at temperatures of _____ $^{\circ}$C or higher.

    A - 150        B - 175        C - 200        D - 225        *E - 250

57. Metasomatism is caused by _____ solutions.

    A - hydromorphic        B - hydrostatic        *C - hydrothermal

    D - polythermal        E - hypothermal

58. Hydrothermal solutions may have _____ dissolved in them.

    A - sodium chloride        B - potassium chloride        C - calcium sulfate

    *D - all of the above        E - none of the above

59. Many hydrothermal mineral deposits are associated with _____ rocks.

    A - deep marine        B - continental shelf        C - massive igneous

    *D - evaporites        E - glacial

60. Blueschist and eclogite facies tend to occur under _____ pressure and _____ temperature.

    A - high/high        *B - high/low        C - low/low        D - low/high

61. When crustal rocks are dragged down by a rapidly subducting plate the minerals are subjected to _____ pressure and _____ temperature.

    A - high/high        *B - high/low        C - low/low        D - low/high

62. Greenschist and amphibolite facies are associated with _____ .

    A - cataclastic metamorphism and contact metamorphism

    *B - contact metamorphism and regional metamorphism

    C - regional metamorphism and burial metamorphism

    D - burial metamorphism and cataclastic metamorphism

## TRUE-FALSE

1. Low-grade metamorphism refers to metamorphic processes occurring at temperatures from about 200$^{\circ}$ to 320$^{\circ}$ C, and at relatively low pressures.

    *T/F

2. High-grade metamorphism refers to metamorphic processes occurring at temperatures greater than 550° C, and at high pressures.

    *T/F

3. Intergranular fluids are pure water.

    T/*F

4. Prograde metamorphic reactions are faster than retrograde metamorphic reactions.

    *T/F

5. Metamorphic reactions are easy to experiment with in the laboratory.

    T/*F

6. High temperature, high pressure, long duration metamorphic reactions produce large crystals.

    *T/F

7. Granulites are well foliated.

    T/*F

8. The most valuable variety of marble pf pure white.

    *T/F

9. Metamorphic aureoles are roughly proportional to the size of the intrusions that form them.

    *T/F

10. Regional metamorphism occurs in zone of tectonic plate collision or subduction.

    *T/F

11. An isograd is a line on a map connecting points of first occurrance of a given mineral in metamorphic rock.

    *T/F

12. Metasomatism is associated with very small quantities of water.

    T/*F

13. Hydrothermal solutions form veins by deposition dissolved constituents in cracks they flow through.

    *T/F

14. Origin of hydrothermal solutions is usually easy to decipher.

    T/*F

15. During prograde metamorphism, mineral assemblages are constant.

## COMPLETION

1. The upper limit of metamorphism is when **melting** occurs.

2. Composite volumes of rock containing an igneous component formed by a small amount of melting plus a metamorphic portion, are called **migmatite** .

3. Intermediate and high grade metamorphic rocks have crystal sizes you can see with **the naked eye**.

4. The presence of **slaty cleavage** is clear proof that a rock has gone from being a sedimentary rock to a metamorphic rock.

5. Burial metamorphism is characterized by the **zeolite** family of minerals.

6. The chemical compositions of most rocks are **little** changed by metamorphism.

7. The principal ingredient of hydrothermal solutions is **water**.

## DISCUSSION

1. Discuss the relationship between the presence/absence of pore fluids and the rates of metamorphism.

2. Discuss the relationships between temperature-pressure-time and grain size during metamorphism.

3. Describe the sequence of rocks that form from a shale as temperature-pressure increases.

4. Describe and discuss regional metamorphism.

5. Discuss the concept of metamorphic zones.

6. Describe and discuss the history and basis for the development of the concept of metamorphic facies.

# CHAPTER 6

## GEOLOGIC TIME

### MULTIPLE CHOICE

1. The person principally responsible for development the utilization of the concept of Uniformitarianism was _____ .

    *A - Lyell     B - Hutton     C - Eicher     D - Smith     E - Halley

2. The first scientist to understand the profound significance of relative time in geology.

    A - Lyell     *B - Hutton     C - Eicher     D - Smith     E - Hall

3. Radioactivity was discovered in _____ .

    A - 1783     B - 1865     *C - 1896     D - 1906     E - 1912

4. Reconstruction of the history of the Earth's history is based upon the study of _____ rocks.

    A - igneous     *B - sedimentary     C - metamorphic

5. The study of strata (layered rocks) is called _____ .

    A - sedimentology     B - paleontology     C - structural geology

    *D - stratigraphy     E - positionism

6. Water-laid sediments are deposited in strata that are _____ .

    *A - horizontal     B - slope steeply     C - vertical     D - undulate

    E - none of the above

7. Which of the following can be used to determine whether sedimentary rocks are right side up or upside down.

    A - ripple marks     B - graded beds     C - cross-stratified beds

    *D - all of the above     E - none of the above

8. Rocks that have been deposited layer after layer without interruption are described as _____ .

    A - continuous     B - contiguous     C - coterminous     D - collateral

    E - conformable

9. A substantial break or gap in a stratigraphic sequence is a(n) _____ .

    A - noncontinuousity     B - discontiguousness     C - noncoterminity

D - discolateralization          *E - unconformity

10. An unconformity records a change in environmental conditions that caused_____ .

A - deposition to cease for a considerable time

B - a major compaction and thickness loss

C - erosion that resulted in loss of part of some earlier-formed strata

D - all of the above          *E - only A and C

11. When older strata were deformed and then truncated by erosion before the younger layers were deposited across them is a(n) _____ .

A - nonconformity          B - sloping unconformity          C - disconformity

*D - angular unconformity    E - graded unconformity

12. An irregular surface of erosion between parallel strata is a(n) _____ .

A - nonconformity          B - sloping unconformity          *C - disconformity

D - angular unconformity    E - graded unconformity

13. The hardest type of unconformity to recognize is a(n) _____ .

A - nonconformity          B - sloping unconformity          *C - disconformity

D - angular unconformity    E - graded unconformity

14. Disconformities are usually recognized through the discovery that ____.

A - minerals in adjacent layers are very different ages

B - rocks above and below are not parallel

C - environments in adjacent layers are very different ages

*D - fossils in adjacent layers are very different ages

E - sedimentary structures above and below are not congruous

15. Strata overlying igneous or metamorphic rocks is a(n) _____ .

*A - nonconformity          B - sloping unconformity          C - disconformity

D - angular unconformity    E - graded unconformity

16. A study of unconformities brings out the close relationship between _____ .

A - tectonic          B - erosion     C - sedimentation          *D - all of the above

E - only B and C

17. Any distinctive rock unit that differs from the strata above and below.

*A - rock stratigraphic unit          B - time-stratigraphic unit     C - system

D - period                    E - formation

18. A group of similar strata that are sufficiently different from adjacent groups of strata so that on the basis of physical properties they constitute a distinctive, recognizable unit that can be used for geologic mapping over a wide area.

*A - rock stratigraphic unit          B - time-stratigraphic unit

C - system                    D - period                    E - formation

19. A unit representing all the rocks that formed during a specific interval of geologic time.

A - rock stratigraphic unit          *B - time-stratigraphic unit

C - system                    D - period                    E - formation

20. Time-stratigraphic units have upper and lower boundaries that are each _____ .

A - horizontal          B - undulating                *C - everywhere the same age

D - all of the above            E - only A and B

21.  The ages of the boundaries of a formation can _____ from place to place.

A - be the same          B - differ          C - transgress time

*D - all of the above          E - only A and B

22. The primary time-stratigraphic unit is a _____ .

A - period     B - era          C - epoch     *D - system     E - stage

23.  System names have _____ applicability.

A - local          B - regional  . C - continental          D - intercontinental

*E - worldwide

24. Most systems are known to encompass absolute time intervals of _____ of years.

A - thousands          B - tens of thousands          C - hundreds of thousands

D - millions          *E - tens of millions

25. The time during which a system accumulated is a _____ .

*A - period     B - era          C - epoch          D - age          E - stage

26. A period is a _____ unit .

A - material          *B - nonmaterial

27. The principle of correlation was developed by _____ .

A - Lyell          B - Hutton          C - Eicher          *D - Smith          E - Hall

28. The determination of equivalence in time-stratigraphic age of the succession of strata found in two or more different areas is _____ .

    A - equivalency        B - coexistance        C - commonness

    *D - correlation        E - determinants

29. For distances of a few to a few tens of kilometers rocks can be correlated using _____ .

    A - physical similarity        B - fossil content

    C - presence of sedimentary structures        D - all of the above    *E - only A and B

30. For distances of a few hundred to a few thousand kilometers rocks can be correlated using _____ .

    A - physical similarity        *B - fossil content

    C - presence of sedimentary structures        D - all of the above    E - only A and B

31. During correlation physical characteristics such as _____ are used.

    A - grain size        B - color        C - sedimentary structures

    *D - all of the above        E - only A and C

32. A thin and generally widespread sedimentary bed with characteristics so distinctive that it can be easily recognized but not confused with any other bed is called a _____ .

    A - master bed        *B - key bed    C - area bed    D - equivalency bed    E - index bed

33. A fossil that can be used to identify and date the stratum in which it is found is a(n) _____ .

    A - master fossil        B - key fossil    C - area fossil        D - equivalency fossil

    *E - index fossil

34. Fossil correlation over large geographical areas generally involves _____ .

    *A - assemblages of fossils        B - as few different types of fossils as possible

    C - a single type of fossil if possible        D - B and C

    E - fossil correlation over large geographical areas is impossible

35. The oldest eon in the geologic column is the _____ .

    *A - Hadean    B - Archean    C - Proterozoic    D - Paleozoic    E - Cryptozoic

36. The eon in the geologic column that contains the oldest rocks and first fossils.

    A - Hadean    *B - Archean    C - Proterozoic    D - Paleozoic    E - Cryptozoic

37. The era in the geologic column that contains the first multicellular fossils.

    A - Hadean    B - Archean    *C - Proterozoic    D - Paleozoic    E - Cryptozoic

38. The youngest era in the geologic column, the one that contains the greatest abundance of fossils.

    A - Hadean    B - Archean    C - Proterozoic    *D - Paleozoic    E - Cryptozoic

39. Eons are divided into _____.

    *A - eras    B - periods    C - epochs    D - stages    E - ages

40. The Paleozoic Era is divided into _____ Periods.

    A - 2    *B - 3    C - 4    D - 5    E - 6

41. The Paleozoic Era is divided into periods based upon _____ .

    A - radioactive dating    B - mineralogy    C - rock types

    D - environments of deposition    *E - fossil content

42. Periods of the Phanerozoic Eon are named for sites in _____ .

    A - Asia and South America    B - South America and North America

    *C - North America and Europe    D - Europe and Africa    E - Africa and Asia

43. In 1889 John Joly calculated the age of the Earth using _____ .

    A - radioactivity    B - sedimentation rates vs. thickness of sedimentary rock column

    C - rates of evolution of animals    D - rate of spreading of the continents

    *E - rate of salt added to oceans vs. amount of salt in the oceans

44. Using the saltiness of the oceans, John Joly calculated the age of the Earth to be _____ million years.

    A - 70    *B - 90    C - 110    D - 1500    E - 4600

45. What was the major flaw in Lord Kelvin's calculations of the age of the Earth?

    *A - no additional heat has been added since the Earth was formed

    B - the Earth is not homogeneous

    C - radioactivity adds far less heat than he calculated

    D - his mathematics was totally flawed

    E - the Earth started cold and is progressively heating

46. The nucleus of most elements contains _____ in addition to protons.

    A - electrons    B - positrons    *C - neutrons    D - neutrinos    E - none of the above

47. Atoms of an element that contain different numbers of neutrons are called _____ .

    A - isopacs    B - isopleths    C - isoheyts    D - isograds    *E - isotopes

48. During radioactive decay, a neutron can be transformed into _____ .

    A - a proton    B - an alpha particle    C - a beta particle    D - A and B    *E - A and C

49. The loss of a gamma ray will _____ the mass number or atomic number of an isotope.

    A - increase        B - decrease        *C - have no effect on

50. The number of decaying parent atoms continuously _____while the number of daughter atoms continuously _____ .

    A - increases/increases        B - increases/decreases

    C - decreases/decreases        *D - decreases/increases

51. Careful study of radioactive isotopes in the laboratory has shown that decay rates are _____ by changes in the chemical and physical environment.

    *A - unaffected        B - only slightly affected    C - directly proportionally affected

    D - inversely proportionally affected    E - no such thing occurs

52. The sum of remaining parent atoms plus the number of daughter atoms is equal to _____ .

    A - the combined ages of the sample    B - rate at which the sample is decaying

    *C - the number of parent atoms the sample started with

    D - all of the above        E - only A and B

53. Potassium-40 changes to Argon-40 by _____ .

    A - alpha particle loss        B - beta particle loss        *C - beta particle gain

    D - both A and B        E - both A and C

54. Argon-40 is not included when potassium-bearing minerals originally form because _____ .

    A - Argon-40 is never present in a magma

    B - Argon-40 is not compatible with the crystal lattice of any mineral

    *C - the magma is too hot for Argon-40 to be trapped

    D - Argon-40 has too short a half-live

    E - the magma is too cold for Argon-40 to be mobile

55. $^{40}K/^{40}Ar$ dating is most successfully applied to _____ rocks.

    A - granitic    B - volcanic    C - pyroclastic    D - all equally well

    *E - only B and C

56. There are _____ naturally occurring radioactive isotopes that predominate in geologic studies.

A - 4          B - 5          *C - 6          D - 7          E - 8

57. Radiometric dating with radiocarbon is unique because _____ .

    A - it has available rate of decay          B - it has a very short half-life

    C - the number of daughter atoms cannot be measured

    D - only A and B          *E - only B and C

58. The proportion of carbon-14 in the atmosphere is _____ .

    A - decreasing          B - increasing          *C - nearly constant

    D - highly variable on a short term cycle          E - variable from place to place

59. During the life-time of an organism, the proportion of radiocarbon _____ .

    A - decreases          B - increases          *C - is nearly constant

    D - is highly variable on a short term cycle          E - variable from place to place

60. After death, the proportion of radiocarbon in an organism is _____ .

    *A - decreases          B - increases          C - is nearly constant

    D - is highly variable on a short term cycle          E - variable from place to place

61. The oldest rocks that have been are _____ billion years.

    A - 0.6          B - 3.6          *C - 3.9          D - 4.1          E - 4.6

62. The oldest isolated mineral grains the have been age dated on Earth are _____ billion years.

    A - 0.6          B - 3.6          C - 3.9          *D - 4.1          E - 4.6

63. The time of origin of Earth and the other members of this solar system is approximately _____ billion years.

    A - 0.6          B - 3.6          C - 3.9          D - 4.1          *E - 4.6

64. At temperatures above the _____ point, the thermal agitation of atoms is such that permanent magnetism is impossible.

    A - Kelvin          B - Becquerel          C - Thermal Conversion

    *D - Curie          E - Thermal Agitation

65. Sedimentary rocks become magnetic when _____ .

    A - they are heated above $580^0$ C and allowed to cool

    B - they have large amounts of magnetite

    *C - flakes of iron bearing minerals are oriented during sedimentation

D - only A and B          E - only A and C

66. Sediment cores recovered from the seafloor can be dated very accurately by using _____ .

A - radiometric age dates          B - fossils          C - magnetic reversals

D - only A and C          *E - only B and C

## TRUE-FALSE

1. When looking at sedimentary rock, the oldest are always on the bottom and youngest on the top.

T/*F

2. Breaks in the sedimentary rock record are common.

*T/F

3. Accumulation of sediments at one place compensates for destruction of rock in another.

*T/F

4. A period is a material unit.

T/*F

5. A system is a material unit.

*T/F

6. The geologic record is complete and without unconformities in most places.

T/*F

7. Continuous exposures of rocks over long distances are not common.

T/*F

8. The best examples of index fossils are swimming or floating organisms.

*T/F

9. The best examples of index fossils are organisms that evolved rapidly and quickly became widely distributed.

*T/F

10. Eons are divided into periods.

T/*F

11. The Geologic Column is standard worldwide.

*T/F

12. All isotopes have the same rate of transformation.

T/*F

13. Most of the isotopes of the chemical elements found in the Earth are stable and not subject to change.

*T/F

14. All of the radioactive isotopes that were ever present in the Earth are still present.

T/*F

15. Radioactive isotopes have a wide range of half-lives.

*T/F

16. The various members of this solar system appear to have formed at different times.

T/*F

## COMPLETION

1. The concept that in any sequence of sedimentary strata, the order in which the strata were deposited is from bottom to top is **the principle of stratigraphic superposition**.

2. A thin and generally widespread sedimentary bed with characteristics so distinctive that it can be easily recognized but not confused with any other bed is called a **key bed** .

## DISCUSSION

1. Discuss the concept(include the assumptions and fallacies) that the age of the Earth could be determined by calculating the total thickness of the sedimentary rocks and dividing it by the thickness of sediment deposited in the sea each year.

2. Discuss the concept of rock correlation.

3. Discuss index fossils.

4. Discuss the three ways that radioactive decay occurs.

5. Discuss the concept of half-life and how it is used for radioactive age dating.

6. Discuss how radiocarbon dating works and why.

7. Discuss how the 4.6 billion years age of the Earth was derived.

8. Discuss how igneous and sedimentary rocks become magnetic.

# CHAPTER 7

## WEATHERING AND SOILS

### MULTIPLE CHOICE

1. Weathering occurs when rock and sediment are exposed to _____ .

    A - air        B - moisture        C - organic matter    *D - all of the above

    E - only A and C

2. Weathering is _____

    A - the chemical alteration of rock and sediment

    B - the mechanical breakdown of rock and sediment

    C - the biological assimilation of rock and sediment

    D - all of the above        *E - only A and B

3. Weathering on rock and sediment progresses _____ .

    A - from the inside out        *B - from the surface downward

    C - from the interior to the surface        D - at uniform rates worldwide

    E - never involves water

4. Chemical weathering produces compounds that are _____ .

    A - less stable near the Earth's surface        B - water soluble

    C - higher in organic matter    *D - more stable near the Earth's surface

    E - isolated from mechanical weathering products

5. Mechanical weathering produces products that are _____ .

    A - less stable near the Earth's surface        B - water soluble

    C - higher in organic matter        D - more stable near the Earth's surface

\*    E - smaller pieces of the parent material

6. When deeply buried rocks are brought near the Earth's surface by erosion, _____ develop because the rocks expand.

    *A - joints    B - faults    C - vesicles    D - dikes    E - sills

7. When sheet-like igneous rock bodies at, or near the surface cool rapidly _____ are commonly produced.

    A - faults       *B - columnar joints        C - hoo doos   D - towering prisms

    E - mass breakouts

8. When water freezes and expands in joints and cavities of rock _____ occurs.

    A - columnar jointing       B - sheeting        * C - frost wedging

    D - unloading         E - thermal expansion

9. Frost wedging is most effective at temperatures of _____ C.

    A - 0° to -10°   B - 0° to -5°   *C - -5° to -15°     D - -10° to -20°     E - -15° to -25°

10. Laboratory experiments with heating and cooling of rocks over the diurnal temperature range has yielded results that _____ .

    A - show it is a major weathering agent    B - indicate it is a minor weathering agent

    *C - show no effect as a weathering agent

11. Fire is a _____ .

    *A - very effective weathering agent    B - weathering agent of little importance

    C - weathering agent of no importance

12. A thin outer shell of material that expands and breaks away from a rock that has been intensely heated is a _____ .

    A - exfoliation       B - sluff layer       C - expansion layer

    D - fire layer       *E - spall

13. Plant root wedging is probably a _____ agent of weathering.

    *A - very important   B - slightly important       C - inconsequential weathering

14. Compared to the site of origin of most minerals, the Earth surface is _____ .

    A - high temperature-high pressure       B - high temperature-low pressure

    C - low temperature-high pressure       *D - low temperature-low pressure

15. Chemical reactions are most effective under _____ temperature- _____ pressure conditions.

    *A - high-high       B - high-low       C - low-high       D - low-low

16. As rainwater falls through the atmosphere, it dissolves small quantities of carbon dioxide, producing _____ .

    A - hydrochloric acid       B - sulfuric acid       C - phosphoric acid

    *D - carbonic acid       E - acetic acid

17. During a hydrolysis reaction, $H^{1+}$ ions replace potassium ions in potassium feldspar and the mineral _____ is formed.

A - plagioclase        B - augite        *C - kaolinite        D - diopside        E - calcite

18. A chemical reaction in which the $H^{1+}$ or $OH^{1-}$ ions in water replace ions of a mineral, is called _____ .

*A - hydrolysis        B - leaching        C - oxidation        D - hydration        E - reduction

19. The continued removal, by water solutions, of soluble matter from bedrock or regolith.

A - hydrolysis        *B - leaching        C - oxidation        D - hydration        E - reduction

20. Soluble substances leached from rocks during weathering are present in all _____ waters.

A - surface        B - ground        C - connate        D - all of the above        *E - only A and B

21. A chemical reaction where oxygen combines with another element is _____ .

A - hydrolysis        B - leaching        *C - oxidation        D - hydration        E - reduction

22. a chemical reaction where water combines into a crystal structure of a mineral is _____ .

A - hydrolysis        B - leaching        C - oxidation        *D - hydration        E - reduction

23. Granite weathers by hydrolysis to produce clay plus _____ .

A - $Na^{1+}$        B - $K^{1+}$        C - $Mg^{2+}$        *D - all of the above        E - none of the above

24. Basalt weathers by hydrolysis to produce clay plus _____ .

A - $Na^{1+}$        B - $Ca^{2+}$        C - $Mg^{2+}$        *D - all of the above        E - none of the above

25. Minerals such as gold, silver, platinum are _____ by chemical weathering.

* A - uneffected        B - slightly effected        C - strongly effected

26. The weathering rind on basalt cobbles is brown because of the contained _____ .

A - hematite        *B - goethite        C - pyrite        D - hornblende        E - olivine

27. During weathering, thin shells of rock may spall off from the outside of an outcrop or a boulder, a process known as _____ .

A - layering        B - shelling        C - sluffing        D - slumping        *E - exfoliation

28. As exfoliation progresses, the rock becomes more _____ .

A - square        B - platy        C - tabular        *D - spherical        E - none of the above

29. During weathering, the volume of the weathered produce becomes _____ than the original unweathered material.

*A - greater        B - less        C - remains constant

30. The effectiveness of chemical weathering _____ as the surface area exposed to weathering increases.

     A - remains constant       B - has no relationship     *C - increases  D - decreases

31. Quartz-rich rocks are _____ to chemical weathering.

     A - susceptible       *B - resistant       C - unpredictable

32. Quartz-rich rocks typically form _____ .

     A - hilly terrain       B - mountainous terrain       C - ridges

     *D - all of the above       E - none of the above

33. Rock on steep slopes is weathered _____ .

     A - deeply       *B - to only shallow depths       C - follows no pattern

34. In a warm, humid climate climate, limestone produces a _____ topography.

     A - bold cliffy       *B - low gentle

35. In a dry climate climate, limestone produces a _____ topography.

     *A - bold cliffy       B - low gentle

36. The part of the regolith that can support rooted plants is _____ .

     A - alluvium  B - fanglomerate     C - tephra     *D - soil     E - lahars

37. Plants derive nutrients from _____ .

     A - nutrients released from decaying organisms     B - magmatic solutions

     C - nutrients released from weathering of minerals     D - all of the above

     *E - only A and C

38. Soil horizons represent _____.

     *A - physical and chemical changes to the regolith     B - layers of sediment

     C - unconformable relationships    D - only A and C     E - only B and C

39. The uppermost soil horizon representing an accumulation of organic matter.

     *A - O horizon     B - A horizon  C - B horizon  D - E horizon     E - C horizon

40. The upper dark colored horizon that has suffered loss of material by downward transportation of clay and leaching of soluble minerals.

     A - O horizon     *B - A horizon     C - B horizon  D - E horizon  E - C horizon

41. A light colored(grayish or whitish) layer below the A horizon that lacks darker colored oxide coatings on light-colored minerals. Commonly found in acidic soils that develop beneath evergreen forests.

A - O horizon          B - A horizon  C - B horizon  *D - E horizon          E - C horizon

42. Horizon commonly brownish or reddish color. enriched in clay and/or iron and aluminum hydroxides produced from weathering.

A - O horizon          B - A horizon  *C - B horizon          D - E horizon  E - C horizon

43. Horizon consisting of parent material in various stages of weathering.

A - O horizon          B - A horizon  C - B horizon          D - E horizon  *E - C horizon

44. Soils formed in cold, high-latitude deserts, soils generally are well drained and lack well-developed horizons.

*A - Entisols  B - Histosols  C - Inceptisols          D - Alifsols  E - Spodosols

45. Soils formed in high-latitudes or high altitude, poorly drained and generally water logged and rich in organic matter.

A - Entisols  *B - Histosols  C - Inceptisols          D - Alifsols  E - Spodosols

46. High latitude-high altitude well-drained soils with poorly developed A and B horizons.

A - Entisols  B - Histosols  *C - Inceptisols          D - Alifsols  E - Spodosols

47. Soils in a temperate climate vary largely in response to differences in _____ .

A - parent material  B - climate  C - vegetation cover  D - A and B  *E - B and C

48. Deciduous woodlands, typically with a clay-rich B horizon beneath a light gray E horizon.

A - Entisols  B - Mollisols  C - Ultisols  *D - Alifsols  E - Spodosols

49. Acidic soils developed in cool, moist evergreen forests with an organic-rich A horizon, an ashlike E horizon, and an iron-rich B horizon.

A - Entisols  B - Mollisols  C - Ultisols  D - Alifsols  *E - Spodosols

50. Soils in mountainous terrains, where a cool climate and steep, eroding slopes maintain low rates of soil development, frequently have minimally developed profiles.

*A - Entisols  B - Mollisols  C - Ultisols  D - Alifsols  E - Spodosols

51. Soils formed in subtropical climates commonly display a strongly weathered B horizon.

A - Entisols  B - Mollisols  *C - Ultisols  D - Alifsols  E - Spodosols

52. Grassland and prairie soils with thick dark-colored, organic-rich A horizons.

A - Entisols  *B - Mollisols  C - Ultisols  D - Alifsols  E - Spodosols

53. An almost impervious layer of whitish calcium carbonate in desert soils.

A - calcsol  B - cretsol  *C - caliche  D - rockpan  E - soil pavement

54. Strongly alkaline desert soil with a well developed K horizon of calcium carbonate.

A - Entisols  B - Mollisols  C - Ultisols  *D - Aridosols  E - Spodosols

55. Soils formed where rainfall is high and the average temperature is very warm, characterized by extreme chemical alteration of the parent material, extremely weathered and infertile because essential nutrients have been leached away.

    A - Entisols   *B - Oxisols   C - Ultisols   D - Vertisols      E - Spodosols

56. Equatorial and tropical soils developed in alternating wet and dry seasons, a high clay content causes the soil to swell and shrink with wetting and drying, develop deep cracks when dry.

    A - Entisols   B - Oxisols   C - Ultisols   *D - Vertisols      E - Spodosols

57. Among the least soluble weathering products are _____ .

    A - quartz   B - hematite   C - goethite   D - A and B   *E - B and C

58. Product formed in very wet and warm tropical regions, rock-forming minerals are leached away, leaving a soft, mottled, reddish-gray residue, rich in iron.

    A - caliche   B - ferrecrete      C - ironcrete   *D - laterite   E - ferrestone

59. A soil that formed at the ground surface and subsequently was buried and preserved.

    A - Entisols   B - Oxisols   C - Ultisols   *D - Paleosols   E - Spodosols

60. Eroded soil ends up _____ .

    A - deposited along valley floors        B - in marine deltas

    C - in reservoirs behind large dams       *D - all of the above   E - only A and B

61. Soil erosion is a problem of catastrophic proportions. It is estimated that for each pound of food produced in the United States _____ pounds of soil is lost.

    A - 2     *B - 5     C - 10     D - 15     E - 25

62. Soil is lost to erosion because _____ .

    A - land that is too steep is farmed        B - too many row crops are planted

    C - soil in too dry climates is farmed       *D - all of the above   E - only A and B

63. Rates of soil erosion can minimized by _____ .

    A - not overgrazing pastures

    B - alternating row crops(corn) with solid cover crops(grass)

    C - minimize slopes by land leveling or terracing   *D - all of the above   E - A and B

64. Lateritic concentrations of ____ are economically important.

    A - iron   B - aluminum      C - nickel   D - all of the above   *E - B and C

65. Secondary enrichment of minerals by weathering has produced important deposits of _____ .

A - iron ore    B - manganese ore    C - copper ore    *D - all of the above

E - only A and B

66. The most easily weathered minerals at the surface of the earth are formed at _____ .

*A - highest temperature-highest pressure    B - highest temperature-lowest pressure

C - at the surface of the Earth    D - there is no preference of site

E - only B and C

67. High rates of chemical weathering and high mountains are related for several reasons.

A - excessive mechanical weathering cause rapid chemical weathering

B - high mountains are mostly igneous rock that weathers rapidly

C - high mountains receive large amounts of precipitation

D - all of the above    *E - only A and C

## TRUE-FALSE

1. Salt crystals growing along within rock cavities or along grain boundaries cause rupturing or disaggregation of rocks.

   *T/F

2. When water freezes and expands in joints and cavities of rock frost wedging occurs.

   *T/F

3. All igneous rocks weather to produce the same products.

   T/*F

4. Weathering rinds form equally well on all rock types.

   T/*F

5. Soils are not the same everywhere.

   *T/F

6. Soil profiles develop more rapidly than chemical weathering takes.

   *T/F

## COMPLETION

1. The chemical alteration and mechanical breakdown of rock and sediment when exposed to air, moisture, and organic matter is **weathering**.

2. Most of the rock debris seen on high mountain slopes is formed by **ice wedging**.

3. A thin outer shell of material that expands and breaks away from a rock that has been intensely heated is a **spall**.

4.As rainwater falls through the atmosphere, it dissolves small quantities of carbon dioxide, producing **carbonic acid** .

5. Sometimes dissolved ions reach concentrations high enough to give the water an **unpleasant taste**.

6. During weathering, thin shells of rock may spall off from the outside of an outcrop or a boulder, a process known as **exfoliation** .

7. The part of the regolith that can support rooted plants is **soil** .

8. Taken together, the soil horizons constitute a **soil profile**.

9. The 6 soil-forming factors are: **climate, vegetation cover, soil organisms, composition of parent material, topography, and time**.

# DISCUSSION

1. Describe what happens during hydrolysis of potassium feldspar.

2. Discuss the chemical weathering of granite and basalt.

3. Discuss how weathering rinds can be used to measure relative ages of sediment bodies.

4. Discuss the relationship between chemical weathering rates and subdivision of minerals

5. Discuss time as a variable in weathering.

6. Describe and discuss the several soil horizons.

7. Discuss the effects of man and agriculture on world soils.

8. Discuss methods of minimizing rates of erosion on agricultural soils.

9. Discuss the relationship between plate tectonics and worldwide weathering rates.

# CHAPTER 8

## MASS-WASTING

### MULTIPLE CHOICE

1. The movement of regolith downslope, solely as a result of the pull of gravity, without the aid of a transporting medium.

    A - erosion         B - weathering         *C - mass-wasting

    D - system movement         E - hill stabilization

2. Downslope movement of material is caused by _____ .

    A - water     B - wind     C - ice     *D - gravity     E - slope of the land

3. Under natural conditions, a hill slope evolves toward an angle that allows the quantity of regolith reaching any point from upslope to be balanced by the quantity that is moving downslope.  This condition is _____ .

    A - balanced-state         B - balanced input-output         C - slope equivalency

    *D - steady-state         E - no such thing occurs

4. The force tending to cause a rock to move downslope is _____ .

    A - shear strength         B - shear factor         *C - shear stress

    D - tangential shear         E - deformational shear

5. The internal resistance of a body to deformation and down slope movement is _____ .

    *A - shear strength         B - shear factor         C - shear stress

    D - tangential shear         E - deformational shear

6. A slope is most stable if the "Shear Strength/Shear Stress" ratio is _____ .

    A - less than 0         B - equal to 0         C - less than 1         D - equal to 1

    *E - greater than 1

7. Shear stress may increase by _____ .

    A - slope steepening by erosion         B - jolting by an earthquake

    C - decay of plant roots         D - all of the above         *E - only A and B

8. Shear strength may decrease by _____ .

    A - weathering         B - jolting by an earthquake         C - decay of plant roots

D - all of the above          *E - only A and C

9. Mass-wasting is most likely to occur _____ .

    A - on a flat beach          B - on a flat pasture          C - on gently rolling hills

    *D - on a steep mountain side          E - an equally good chance on any of them

10. A small amount of water in regolith tends to _____ its shear strength because of _____ .

    A - decrease/gravitational attraction          B - decrease/capillary attraction

    C - increase/gravitational attraction          *D - increase/capillary attraction

    E - no change occurs with the addition of small amounts of water

11. If too much water is added to regolith the overlying grains are _____ apart.

    A - pried          B - floated          C - hydrauliced          *D - bouyed          E - hydroplaned

12. Water can be instrumental in reducing shear strength and thereby promoting movement of rock and sediment downslope under the pull of gravity by _____ .

    A - reducing the natural cohesiveness between grains

    B - reducing friction at the base through increased fluid pressure

    C - decreasing gravitational attraction of the particles

    D - all of the above          *E - only A and B

13. Mass-wasting processes all share one thing in common: _____ .

    A - they are fast          B - they are very slow          C - they involve water

    *D - they take place on slopes          E - they are easily classified

14. Downward and outward rotational movement of rock or regolith along a curved concave-up surface.  Most common on oversteepened banks.

    *A - slump          B - rockfall          C - debris fall          D - rock slide          E - debris slide

15. Slumps are frequently associated with _____ .

    A - heavy rains          B - oversteepened banks          C - sudden shocks, i.e. earthquakes

    *D - all of the above          E - none of the above

16. Free falling rock bodies.

    A - slump          *B - rockfall          C - debris fall          D - rock slide          E - debris slide

17. With greater distance of fall the _____ the speed reached by a rock.

    *A - greater          B - lesser          C - unrelated

18. Falling rock material stops moving because of _____ .

*A - friction          B - increasing slope    C - decrease of gravitational attraction

D - all of the above          E - only A and C

19. A mass of falling rock, weathered regolith, and vegetation.

A - slump       B - rockfall    *C - debris fall       D - rock slide       E - debris slide

20. The rapid downslope movement of a mass of rock along a sloping surface such as a bedding plane.

A - slump       B - rockfall    C - debris fall       *D - rock slide       E - debris slide

21. The rapid downslope movement of rock, weathered regolith, and vegetation.

A - slump       B - rockfall    C - debris fall       D - rock slide       *E - debris slide

22. The body of debris sloping outward from the cliff that supplies it.

*A - talus       B - fallout       C - basefall       D - foot accumulation

E - base accumulation

23. The steepest stable angle at which fallen fragments come to rest.

A - angle of accumulation          *B - angle of repose       C - slope of repose

D - angle of fall          E - angle of stability

24. The steepest stable angle that loose material will accumulate.

A - 25° to 29°   B - 29° to 33°   *C - 33° to 37°   D - 37° to 41°   E - 41° to 45°

25. Sediment flows involve _____ .

A - water       B - sediment   C - air       *D - all of the above   E - only A and B

26. A sediment flow composed of a mass of water-saturated sediment.

A - granular flow       *B - slurry flow       C - solifluction       D - debris flow

E - mudflow

27. A sediment flow that is a mixture of sediment, water, and air that is grain supported by grain to grain contact.

*A - granular flow       B - slurry flow       C - solifluction       D - debris flow

E - mudflow

28. A very slow type of granular flow, measured in millimeters or centimeters per year.

*A - creep       B - debris avalanche       C - solifluction       D - debris flow

E - mudflow

29. A more rapid type of granular flow, measured in kilometers per year.

    A - creep       *B - debris avalanche       C - solifluction       D - debris flow

    E - mudflow

30. The very slow downslope movement of saturated soil and regolith.

    A - creep       B - debris avalanche       *C - solifluction       D - debris flow

    E - mudflow

31. The downslope movement of unconsolidated regolith, the greater part being coarser that sand, at rates ranging from only about 1 m/yr to as much as 100 km/hr.

    A - creep       B - debris avalanche       C - solifluction       *D - debris flow

    E - mudflow

32. A highly fluid, rapidly moving debris flow.

    A - creep       B - debris avalanche       C - solifluction       D - debris flow

    *E - mudflow

33. One of the major hazards associated with volcanic eruptions is a _____ .

    A - creep       B - debris avalanche       C - solifluction       D - debris flow

    *E - mudflow

34. Rates of creep vary with _____ .

    A - steepness of slope       B - water content       C - amount and type of vegetation

    *D - all of the above       E - only A and B

35. Colluvium can be distinguished for fluid laid sediment by _____ .

    A - angular fragments       B - well layered       C - poorly sorted

    D - all of the above       *E - only A and C

36. Granular flow composed of weak or weathered regolith in sizes of clay and silt at least intermittently saturated with water.

    *A - earthflow       B - liquifaction       C - grain flow       D - debris avalanche

    E - solifluction

37. Rapid fluidization of wet, highly porous clay to sand sized sediments caused by an abrupt shock.

    A - earthflow       *B - liquifaction       C - grain flow       D - debris avalanche

    E - solifluction

38. Rapid downslope movement of dry granular sediment.

A - earthflow          B - liquifaction          *C - grain flow          D - debris avalanche

E - solifluction

39. Debris avalanches can be expected on _____ .

A - flanks of shield volcanos          B - flanks of cinder cones

*C - flanks of stratovolcanos          D - all of the above          E - only A and B

40. From the standpoint of lives lost _____ are probably the most destructive.

A - earthflows          B - liquifactions          C - grain flows

*D - debris avalanches          E - solifluction

41. Mass-wasting is especially active at _____ .

A - high altitudes          B - high latitudes          C - high temperatures

*D - only A and B          E - only A and C

42. The lifting of regolith by the freezing of contained water is _____ .

*A - frost heaving          B - frost wedging          C - frost lifting

D - expansion heaving          E - surface lifting

43. Downslope movement of the thin, thawed surface layer of regolith in cold regions that are underlain by permanently frozen ground.

A - solifluction          B - debris flow          *C - gelifluction          D - liquifaction

E - detachment

44. Submarine slope failures generally do not occur on slopes less than _____ degrees.

A - 12          B - 16          *C - 20          D - 24          E - 28

45. Most of the submarine slump masses and debris flows on the worldwide ocean bottoms accumulated during _____ .

A - the recent stillstand of the oceans          *B - the last glacial age

C - splitting apart of the ocean basin          D - formation of the oceanic ridges

E - major volcanic events

46. Mass-wasting might be expected in material that is _____ .

A - well jointed          B - permeable over impermeable

C - unconsolidated          *D - all of the above          E - none of the above

47. Maps showing areas of potential mass-wasting impact are constructed by _____ .

245

A - studying the bedrock geology        B - mapping sites of former similar events

C - looking for other types of geologic hazards        *D - all of the above    E - A and B

48. Mass-wasting maps of the Cascade Range of northwestern U.S. show that risk from mudflows extends as far as _____ miles from the volcanic slopes.

A - 10            B - 25            C - 50            D - 100            *E - more than 100

49. Most of the world's major historic and prehistoric landslides are clustered along _____ .

A - diverging plate boundaries        B - near hot spots

*C - converging plate boundaries      D - all of the above        E - only A and B

50. Mass-wasting may be more effective as an agent of gradation in settings such as _____ .

A - coastal plains      B - continental interiors      C - continental shield areas

*D - high collision/volcanic mountains        E - they are all equally effected

## TRUE-FALSE

1. Warnings by geologists of impending geologically related disasters often go unheeded by government officials.

*T/F

2. Mass-wasting occurs exclusively on the lands.

T/*F

3. There is a continuum from clear flowing water to a mass of sediment that is dry.

*T/F

4. The greater the distance of fall of a rock the slower its final speed.

T/*F

5. Slurry flows produce well sorted sediments.

T/*F

6. A mudflow is a rapidly moving debris flow.

*T/F

7. Creep occurs on all sloping land.

*T/F

8. Large debris avalanches are relatively common and easy to study.

T/*F

9. Subaqueous mass-wasting appears to be relatively common.

*T/F

10. It is impossible to do anything to mitigate mass-wasting hazards.

T/*F

## COMPLETION

1. If too much water is added to regolith it becomes subject to **failure**.

2. A body of debris sloping outward from the cliff that supplies it is **talus**.

3. A very slow type of granular flow, measured in millimeters or centimeters per year is **creep**.

4. A more rapid type of granular flow, measured in kilometers per year is **debris avalanche**.

5. From the standpoint of lives lost **debris avalanches** are probably the most destructive.

## DISCUSSION

1. Discuss the role water plays in mass-wasting.

2. Discuss the factors controlling flow.

3. Distinguish between slurry flow and granular flow.

4. Discuss slurry flows.

5. Discuss granular flows.

6. Discuss mudflows related to volcanos.

7. Discuss the relationship between frost heaving and downslope creep.

8. Discuss methods of mitigation of mass-wasting hazards.

9. Discuss mass-wasting in relation to plate tectonics.

# CHAPTER 9

## STREAMS AND DRAINAGE SYSTEMS

### MULTIPLE CHOICE

1. A body of water that flows downslope along a clearly defined natural passageway.

    A - stream    B - river    C - creek    D - brook    *E - they are all synonyms

2. The passageway where a stream flows is a _____ .

    A - course    B - path    *C - channel    D - route    E - bed

3. The quantity of water passing a point on the stream bank in a given interval of time is a measure of the stream's _____ .

    A - load    *B - discharge    C - size    D - mass    E - strength

4. The amount that a stream channel falls per unit of horizontal travel is _____ .

    A - rate of fall    B - drop rate    C - slope    *D - gradient    E - gradualism

5. The average gradient of a stream _____ from head to mouth.

    A - increases    *B - decreases    C - remains constant

6. The portion of the annual precipitation that flows back into the oceans across the surface of the land is _____ .

    A - overland flow    B - back flow    *C - runoff    D - streamflow    E - surface flow

7. During a heavy rain the water that initially tends to move down slope in broad, thin sheets, is called _____ .

    *A - overland flow    B - back flow    C - runoff    D - streamflow

    E - surface flow

8. The formula for stream discharge is:  Discharge = _____ .

    A - width X depth    B - width X velocity    C - depth X velocity

    *D - cross-section area X velocity    E - none of the above

9. If discharge increases, but cross-section area is constant, velocity must _____ .

    *A - increase    B - decrease    C - remain constant

248

10. Traveling down a stream from its head to its mouth, we can see that orderly adjustments occur along the channel: discharge _____ .

    *A - increases        B - decreases        C - remains constant

11. Traveling down a stream from its head to its mouth, we can see that orderly adjustments occur along the channel: channel cross-sectional area _____ .

    *A - increases        B - decreases        C - remains constant

12. Traveling down a stream from its head to its mouth, we can see that orderly adjustments occur along the channel: velocity _____ .

    *A - increases        B - decreases        C - remains constant

13. A graph that plots stream discharge against time is a _____ .

    A - barograph        B - pneumograph    *C - hydrograph    D - hygrograph

    E - thermograph

14. As velocity of water increases during a flood, the stream is able to move greater amounts of material and _____ size material.

    *A - larger        B - smaller        C - only the same

15. Flood-frequency curves are useful for flood prediction, except there are a few problems such as _____ .

    A - they are totally inaccurate        B - they reflect only recent past conditions

    C - when climatic conditions change they are no longer valid

    D - all of the above        *E - only B and C

16. The limiting level below which a stream cannot erode the land is called the _____ of the stream.

    A - lower limit        B - lower surface        C - bottomingout    *D - base level

    E - base surface

17. Base level for most streams is _____ .

    A - unpredictable        B - different than any other stream    *C - global sea level

    D - controlled by the discharge        E - seasonally variable

18. Sea level changes over geologically long periods of time due to _____ .

    A - changes in the shape of the ocean basins

    B - changes in the capacity of ocean basins

    C - changes in the volume of continental glaciers

    *D - all of the above        E - only A and B

19. Natural dams on a stream may be caused by _____ .

    A - landslide sediments    B - stream deposits    C - lava flows

    *D - all of the above    E - only A and B

20. Natural dams are _____ features.

    A - permanent    *B - temporary    C - indeterminate

21. Artificial dams are constructed to provide _____ .

    A - flood control    B - water storage    C - hydroelectric power

    *D - all of the above    E - none of the above

22. Artificial dams have finite lifetimes because _____ .

    A - they often break    B - the streams change channels often

    *C - the reservoir behind them fills with sediment

    D - all of the above    E - only B and C

23. Hydroelectric power is really recycled _____ .

    A - solar energy    B - gravitational energy    C - chemical energy

    D - all of the above    *E - only A and B

24. A stream channel with many curves is refereed to as _____ .

    A - winding    B - curvey    C - sigmoidal

    *D - sinuous    E - irregular

25. Where the deepest water in a channel lies on one side, a _____ will develop on the other side.

    *A - bar    B - lip    C - shelf    D - slough    E - reef

26. Bars in a stream develop where water velocity is _____ .

    A - greatest    *B - least    C - average

27. One of a series of smooth, repetitive bends on a river is a _____ .

    A - oxbow    B - bend    C - loop    *D - meander    E - hydraulic dampener

28. The meandering pattern of a stream reflects the way in which a river _____ as uniformly as possible along its course.

    A - minimizes resistance to flow    B - adjusts its gradient

    C - dissipates energy    D - all of the above    *E - only A and C

29. Lowest stream velocity on a stream is _____ .

A - near the bed                    B - near the surface at mid channel

C - toward the outside of the channel        D - along the side        *E - A and C

30. Greatest velocity on a straight segment of a stream is _____ .

A - near the bed                    *B - near the surface at mid channel

C - toward the outside of the channel            D - along the side

E - A and C

31. Greatest velocity as a stream goes around a bend is _____ .

*A - toward the outside of the channel near the surface        B - near the bed

C - near the surface at mid channel        D - along the side        E - A and C

32. On a stream meander, maximum erosion is on the _____ of the meander.

A - inside                *B - outside                C - variable

33. On a stream meander, maximum deposition is on the _____ of the meander.

*A - inside                B - outside                C - variable

34. Deposits of sediment on the inside of a meander is called a _____ .

A - draw bar    B - center bar    *C - point bar        D - curved bar        E - hooked bar

35. Over long periods of time meanders migrate _____ a valley.

A - from side to side in        *B - down        C - up    D - no where    E - only A and C

36. If a migrating meander stops moving, but the one upstream keeps migrating and erodes through into the lower stopped meander, the cutoff meander is called a(n)_____ .

A - abandoned lake            B - meander lake        C - loop lake

*D - oxbow lake                E - hook lake

37. When one meander is abandoned along a stream, the others _____ .

*A - get longer to compensate        B - get shorter to compensate.

C - remain the same length    D - meanders have no relationship to each other

E - none of the above

38. A braided stream pattern tends to develop ib streams with _____ .

A - highly variable discharge                B - easily erodible banks

C - steep gradient        D - all of the above        *E - only A and B

39. At any moment in time only _____% of the channel width of a large braided stream is covered by water.

A - 5        *B - 10        C - 15        D - 20        E - 25

40. Before streams form erosion begins.

    A - rilling      B - sheet erosion      C - raindrop impact    D - all of the above

    *E - only B and C

41. Stream flow characterized by water particles traveling in smooth parallel layers.

    A - turbulent flow            B - smooth flow              C - base flow

    *D - laminar flow             E - undisrupted flow

42. Stream flow characterized by water particles traveling in erratic and complex pattern,
giving rise to swirls and eddies.

    *A - turbulent flow           B - smooth flow              C - base flow

    D - laminar flow              E - undisrupted flow

43. The ability of a stream to pick up particles of sediment from its channel and move them
along depends largely on _____ of the water.

    A - volume                    B - velocity                 C - turbulence

    D - all of the above          *E - only B and C

44. Any detrital sediment deposited by a stream is _____ .

    A - colluvium                 B - deluvium           *C - alluvium

    D - all of the above          E - only A and B

45. The dissolved load of a stream is the result of _____ .

    *A - chemical weathering           B - physical weathering

    C - mechanical weathering    D - scouring           E - mass-wasting

46. Bed load moves by _____ .

    A - rolling      B - sliding      C - saltation    *D - all of the above    E - only A and B

47. Size sediment transported and stream velocity are _____ .

    *A - directly related    B - inversely related           C - unrelated

48. Placer mineral deposits accumulate where there is enough energy to wash _____ minerals,
but not enough energy to wash _____ minerals.

    A - heavy/ light                   *B - light/ heavy

49. The sediment load of a stream is derived from _____ .

    A - inorganic precipitates in water  *B - fine grained regolith washed into the stream

    C - sediment reworked from the streams own banks          D - only A and B

E - only B and C

50. The great bulk of the solution load of a stream consists of only _____ ions.

    A - 4        B - 5        C - 6        *D - 7        E - 8

51. The size of sediment transported by a stream decreases downstream due to _____ .

    A - abrasion of sediment along the course    B - impact of sediment along the course

    C - sorting of sediment by streams        *D - all of the above    E - only A and B

52. The amount of sediment eroded from the land by runoff and transported by streams is _____ .

    A - sediment load        B - base yield  C - sediment volume

    *D - sediment yield        E - yield point

53. Sediment yield appears to be minimal in which climate?

    A - desert        B - grassland        C - prairie

    *D - temperate        E - tropical

54. Streams may lose energy because of a change in _____ .

    A - gradient        B - velocity        C - discharge

    *D - all of the above    E - none of the above

55. The broad flat portion of the valley adjacent to a stream that is inundated by floodwater.

    *A - floodplain        B - natural levee        C -terrace

    D - delta        E - alluvial fan

56. A broad, low ridge of fine alluvium built along the side of a channel by debris-laden flood water.

    A - floodplain        *B - natural levee        C -terrace

    D - delta        E - alluvial fan

57. Typically the natural levee of a large river is constructed of _____ .

    A - clay and fine silt   *B - fine sand and coarse silt        C - coarse sand

    D - gravel and sand        E- mixed sizes from boulders down

58. Typically the floodplain of a large river is constructed of _____ .

    *A - clay and fine silt  B - fine sand and coarse silt        C - coarse sand

    D - gravel and sand        E- mixed sizes from boulders down

59. A remnant of an abandoned floodplain, but may be underlain by bedrock.

A - floodplain                    B - natural levee                 *C -terrace

D - delta                         E - alluvial fan

60. A fan-shaped body of alluvium typically built where a stream leaves a steep mountain valley.

A - floodplain                    B - natural levee                 C -terrace

D - delta                         *E - alluvial fan

61. A sedimentary deposit that forms where a stream flows into a standing water body.

A - floodplain                    B - natural levee                 C -terrace

*D - delta                      E - alluvial fan

62. The coarsest, thick sloping part of a depositional layer on the front of a delta is a ____ layer.

A - topset     B - frontset     *C - foreset     D - bottomset     E - midset

63. The finer, thinner layer of sediment at the bottom of the delta is a ____ layer.

A - topset     B - frontset     C - foreset     *D - bottomset     E - midset

64. Coarse channel deposits with finer intrachannel sediments covering the remainder of the delta is the ____ layer.

*A - topset     B - frontset     C - foreset     D - bottomset     E - midset

65. The total area that supplies water to a stream.

A - supply area               B - source area              C - drainage area

*D - drainage basin           E - supply basin

66. The line that separates adjacent drainage basins.

*A - divide     B - ridge     C - separation line     D - cap     E - edge

67. When studying a map, the smallest segments that lack are ____ order.

A - primary     B - beginning     *C - first     D - highest     E - none of the above

68. As stream order increases, the number of segments of a particular order ____ .

A - increases     *B - deceases     C - there is no change in abundance

69. The interception and diversion of one stream by another stream that is expanding its basin by erosion in a headward direction is ____ .

A -stream takeover           *B - stream capture           C - stream piracy

D - basin capture           E - basin piracy

70. The ease with which a formation is eroded by streams depends chiefly on its _____ .

    A - thickness             B - composition             C - structure

    D - all of the above           *E - only B and C

71. The line separating two large regions in which the through flowing streams empty into different oceans.

    A - ocean divides           B - basin divides           C - river divides

    *D - continental divides       E - separation lines

72. The Arctic Ocean-Atlantic Ocean Continental Divide was determined largely by _____ .

    *A - continental glaciation         B - an ancient east-west mountain range

    C - the site of major plate tectonic collisions      D - all of the above

    E - only B and C

72. The Gulf of Mexico-Atlantic Ocean Divide was determined largely by _____ .

    A - continental glaciation         B - an ancient north-south mountain range

    C - the site of major plate tectonic collisions      D - all of the above

    *E - only B and C

## TRUE-FALSE

1. The relationship between a stream and its channel if constant.

    T/*F

2. Over time, meanders migrate slowly down a valley.

    *T/F

3. All of the meanders in a river migrate at the same, constant rate.

    T/*F

4. Laminar flow is the pattern to be expected in most streams.

    T/*F

5. Streams that receive most of their water from underground sources generally have a higher concentration of dissolved ions than those whose water comes mainly from surface runoff.

    *T/F

6. Greatest local sediment yields are from landscapes transitional between full desert conditions and grassland.

    *T/F

7. The history of a stream valley is generally very simple.

   T/*F

8. Deltas are highly variable in structure and morphology.

   *T/F

9. A drainage system is a dynamic system tending toward a condition of equilibrium.

   *T/F

10. The study of topographic maps yields virtually no information concerning the underlying geology.

    T/*F

## COMPLETION

1. A stream channel is so responsive to changes in discharge that the system, at any point along the stream, is always close to a **balanced condition**.

2. The measure of how often a flood of a given magnitude is likely to recur is called the **recurrence interval**.

3. Base level for most streams is **global sea level**.

4. Hydropower energy is a **renewable** resource.

5. When a stream loses **energy**, its transport power drops and it deposits part of its load.

## DISCUSSION

1. List and discuss the 4 functions, outlined in the text, that streams serve as geological agents.

2. Discuss the evidence that J Harlen Bretz found to support the idea that the Channeled Scablands of Washington State were water sculptured and how he reached his final conclusions.

3. Discuss uses for a large dam-reservoir complex that has been completely silted-up.

4. Discuss the structure and character of braided streams.

5. Discuss the movement of material as bedload.

6. Discuss the several factors that control sediment yield and how they interrelate.

7. Discuss the sedimentation and history of construction of an alluvial fan.

8. discuss the relationship between drainage patterns, rock structure, and stream history.

9. Discuss how geologic history and plate tectonics have controlled the sites of the present continental divides in North America.

# CHAPTER 10

## GROUNDWATER

### MULTIPLE CHOICE

1. The volume of groundwater is approximately _____ times the volume of water contained on all of the world's freshwater lakes and flowing streams.

    A - 10        B - 20        C - 30        *D - 40        E - 50

2. Most groundwater originates as _____ .

    A - desalinated infiltration from oceans    B - leakage from the interior of the Earth

    *C - rainfall        D - all of the above        E - only A and B

3. The groundwater system is _____ .

    A - nonmoving        B - moving through the ground toward the oceans

    C - moving toward the surface where it runs into the streams

    D - moving toward the interior to become part of igneous rocks

\*    E - B and C

4. Most of the useable groundwater is within approximately _____ m of the surface.

    A - 200        B - 350        C - 500        *D - 750        E - 1000

5. Relatively little available groundwater lies below 750 m because _____ .

    A - gravity is less effective at pulling water that deep

\*    B - pore space diminishes due to great overlying pressure

    C - it is too deep to drill holes effectively

    D - it is so hot at that depth the water turns to stream

    E - all 4 of these are ridiculous choices

6. The portion of the subsurface where the pores have some water but contain mostly air.

    A - zone of aeration        B - pneumatic zone        C - unsaturated zone

    D - all of the above        *E - only A and C

7. The portion of the subsurface where the pores are continuously filled with water.

    *A - saturated zone        B - filled zone        C - nonpneumatic zone

D - all of the above          E - only A and C

8. The boundary between the zones of aeration and saturation.

    A - boundary zone          B - saturation surface          *C - water table

    D - water surface          E - water saturation boundary

9. The water table is maintained at normal levels by _____ .

    A - desalinated infiltration from oceans     B - leakage from the interior of the Earth

    *C - rainfall infiltrating the ground        D - dehydration of minerals

    E - lateral migration of other groundwater

10. Groundwater typically moves at rates of _____ .

    A - kilometers per hour          B - meters per hour          C - kilometers per day

    D - meters per day          *E - centimeters per day

11. The percentage of the total volume of a body of regolith or bedrock that consists of open spaces _____ .

    A - voidality          B - emptyness          C - permeability          *D - porosity

    E - openness

12. The porosity of rock or regolith is affected by _____ .

    A - size of particles          B - shape of particles

    C - compactness of the arrangement of the particles          *D - all of the above

    E - only A and B

13. The porosity of rock or regolith is affected by _____ .

    A - amount of cement in the rock          B - volume of joints in the rock

    C - compactness of the arrangement of the particles

    *D - all of the above          E - only A and B

14. A measure of how easily a solid allows fluids to pass through it is _____ .

    A - fluidity     *B - permeability     C - transference     D - through passage

    E - both A and C

15. Permeability is affected by _____ .

    A - size of pores          B - shape of pores          C - continuity of pores

    *D - all of the above          E - only A and B

16. Soil is usually _____ permeable than the underlying regolith.

    A - more                *B - less

17. Water percolates from areas where the water table is _____ toward areas where it is _____ .

    A - lowest/highest    *B - highest/lowest    C - in largest pores/in smallest pores

    D - in smallest pores/in largest pores        E - there is no such activity as percolation

18. New water enters the groundwater system through the process of ____ .

    *A - recharge           B - replenishment       C - infusion   D - diffusion  E - soaking

19. Discharge of groundwater occurs _____ .

    A - in streams        B - in swamps        C - in lakes    *D - all of the above

    E - only A and C

20. In Darcy's Law the letter "K" the coefficient of permeability stands for _____ .

    A - acceleration due to gravity      B - permeability      C - viscosity of water

    *D - all of the above              E - only B and C

21. People generally obtain supplies of groundwater from _____ .

    A - springs         B - excavated wells    C - lake bottoms where seepage occurs

    D - all of the above          *E - only A and B

22. Simple springs occur where _____ .

    A - lakes cover the water table           B - streams act as discharge areas

    *C - the land surface intersects the water table       D - the water table is unusually deep

    E - the water table has been excavated artificially

23. Springs are often associated with aquicludes such as _____ .

    A - permeable sand overlying impermeable clay intersects the surface

    B - the trace of a fault intersects the surface

    C - jointed lava bed on top of a clay lake bed intersects the surface

    D - all of the above        *E - only A and C

24. The cone of depression for a well is stable if _____ .

    A - rate of withdrawal is constant       B - rate of recharge is constant

    C - new cone of depression does not intersect it    *D - all of the above    E - B and C

25. A water table may be lowered artificially by _____ .

*A - excessive withdrawal of water from wells      B - lack of recharge

C - excessive recharge      D - new springs appearing      E - mass-wasting

26. A body of highly permeably rock or regolith lying in the zone of saturation is a _____ .

   A - aquiclude      B - aquanaut      *C - aquifer   D - aquarium E - aquivour

27. An unconfined aquifer is _____ .

   A - between 2 aquicludes      B - above the general water table and separate from it

   *C - one with the water table open to the surface      D - all of the above      E - A and B

28. An confined aquifer is _____ .

   *A - between 2 aquicludes      B - above the general water table and separate from it

   C - one with the water table open to the surface      D - all of the above      E - A and B

29. An artesian aquifer will _____ .

   *A- cause water to rise in a well above where it is first hit

   B - have an unusually large cone of depression

   C - always cause water to come to the surface without pumping

   D - occur only in France      E - always run out of water very rapidly

30. When ground water is removed fasted than it recharges _____ .

   A - it is being mined      B - it is a nonrenewable resource

   C - it can easily be remedied by modern technology

   D- all of the above      *E - only A and B

31. Lowering of a water table will cause _____ .

   A - drying-up of springs and streams      B - wells to dry-up

   C - infiltration from urban area runoff to be slowed      D - all of the above

   *E - only A and B

32. Artificial recharge may be accomplished by _____ .

*      A - channeling runoff from urban areas into infiltration basins

   B - drilling shallower wells      C - increasing annual precipitation

   D - using less water      E - applying modern technology to well drilling

33. Hard water typically comes from _____ .

   A - sandstone      *B - limestone      C - shale      D - slate      E - lava

34. Soft water typically comes from _____ .

    A - marble     B - dolostone       C - limestone   *D - lava beds   E - carbonatite

35. You get a much nicer lather using soap and _____ water.

    A - hard       *B - soft

36. Biologically contaminated water can be fairly easily purified by _____ .

    A - percolating it through coarse gravel     *B - percolating it through fine sand

    C - allowing it travel through major joint systems in aquifers

    D - allowing the water table to be lowered

    E - by introducing germicides into the groundwater system

37. Municipal landfills are an ideal method to dispose of virtually any sort of waste because _____ .

    A - they are always carefully sited originally

    B - they are carefully buried and safe from ground water percolation

    C - buried materials stabilize quickly     D - all of the above

    *E - none of the above

38. Contaminated groundwater is easily cleaned up by _____ .

    A - allowing it to travel through cavernous limestone

    B - leaving it in the ground more than 5 years

    C - accelerating artificial recharge rates     D - all of the above     *E - none of the above

39. Hazards from underground disposal of nuclear wastes may be minimized by _____ .

    A - selecting sites unlikely to be volcanic     B - selecting sites that are impermeable

    C - selecting sites with no economic value either present or future   *D - all of the above

    E - there is no safe way to dispose of nuclear wastes.

40. As rainwater infiltrates the ground, it begins to react with minerals in the rock and regolith in the chemical reaction _____ .

    *A - dissolution     B - hydration   C - hydrolysis     D - oxidation   E - reduction

41. The most intensely attacked rocks by groundwater solutions are _____ .

    A - granite, rhyolite, andesite       B - slate, phyllite, schist

    *C - limestone, dolostone, marble       D - gneiss, quartzite, hornfels

    E - basalt, gabbro, diorite

42. In a humid climate the greatest rate of removal of surface material is by _____ .

    A - sheet erosion of regolith       *B - dissolution of carbonates

    C - mass-wasting       D - stream erosion       E - rilling of regolith

43. The most common cementing agents in sedimentary rocks are _____ .

    *A - calcite, quartz, iron compounds       B - pyrite, quartz, gypsum

    C - gypsum, halite, muscovite       D - augite, pyroxene, olivine

    E - calcite, gypsum, augite

44. Most caves form by dissolution by _____ acid.

    A - sulfuric    B - hydrochloric    C - phosphoric    *D - carbonic  E - nitric

45. Dripstone and flowstone in caves are composed primarily of _____ .

    A - calcium sulfate    B - silica dioxide    C - iron oxide  *D - calcium carbonate

    E - iron carbonate

46. Dripstone forms when a drop of water loses _____ and precipitates a molecule of calcium carbonate.

    A - oxygen    *B - carbon dioxide    C - nitrogen trioxide  D - methane

    E - hydrogen sulfide

47. A large dissolution cavity that is open to the sky is a _____ .

    A - surface cave    B - surface dissolution chamber    C - sinking chamber

    D - cratered chamber    *E - sinkhole

48. Sinkholes appear at the surface _____ .

    *A - suddenly and unexpectedly    B - slowly and with ample warning  C - only rarely

    D - in all types of bedrock and regolith    E - most commonly in sandstone

49. Karst regions are most commonly underlain by _____ .

    A - lava flows    B - granite batholiths    C - massive sandstone

    *D - massive carbonates    E - shale beds

50. Karst regions are distributed _____ .

    A - only in the tropics    B - only in deserts    C - only in temperate climates

    D - commonly in polar climates    *E - worldwide

## TRUE-FALSE

1. The top of the water table is flat and featureless.

   T/*F

2. A rock with low porosity is likely to have low permeability.

   *T/F

3. Most of the groundwater entering a stream travels along shallow paths not far beneath the water table.

   *T/F

4. Groundwater flows at a constant rate everywhere.

   T/*F

5. Igneous rocks never produce groundwater.

   T/*F

6. Ground water is always safe from contamination.

   T/*F

7. Soft water makes much better soap lather than hard water.

   *T/F

8. Incursion by seawater into freshwater supplies is very difficult to reverse.

   *T/F

## COMPLETION

1. The top of the zone of aeration is the **water table**.

2. A well will supply water if it **intersects the water table**.

3. The most common source of pollution in wells is **sewage**.

4. The conversion of sediment into sedimentary rock is primarily the work of **groundwater**.

## DISCUSSION

1. Discuss the relationship between pore size and molecular attraction.

2. Contrast recharge areas in humid and arid regions.

3. Describe a perched water table.

4. Discuss the structure of the High Plains aquifer, its use and its future.

5. Describe the Floridian Aquifer.

6. Discuss water as a nonrenewable resource.

7. Discuss the relationship between usefulness of groundwater and its type and amount of dissolved mineral matter.

8. Discuss the sequence of events involved in the formation of caves.

# CHAPTER 11

## GLACIERS AND GLACIATION

### MULTIPLE CHOICE

1. A glacier occupying a protected bowl-shaped depression on a mountainside, open downward and bounded upslope by a steep cliff.

    \*A - cirque glacier        B - valley glacier        C - fjord glacier

    D - piedmont glacier        E - ice cap

2. A glacier occupying the gently sloping area in front of a mountain front.

    A - cirque glacier        B - valley glacier        C - fjord glacier

    \*D - piedmont glacier        E - ice cap

3. A glacier occupying a deep valley along a coast.

    A - cirque glacier        B - valley glacier        \*C - fjord glacier

    D - piedmont glacier        E - ice cap

4. A mass of ice covering a mountain highland or lower-lying land at high latitude, and displays generally radial outward flow.

    A - cirque glacier        B - valley glacier        C - fjord glacier

    D - piedmont glacier        \*E - ice cap

5. A continent-sized mass of ice that overwhelms nearly all the land surface within its margin.

    \*A - ice sheet        B - ice shelf        C - fjord glacier

    D - piedmont glacier        E - ice cap

6. A thick, nearly flat sheet of floating ice that is fed by one or more glaciers on land.

    A - ice sheet        \*B - ice shelf        C - fjord glacier

    D - piedmont glacier        E - ice cap

7. The location of the snowline is controlled largely by _____ .

    A - elevation        B - temperature        C - precipitation

    D - all of the above        \*E - only B and C

8. As glacier ice is buried to ever greater depths the size of the crystals _____ .

A - decreases                    B - remains the same              *C - increases

9. On a worldwide basis, glaciers seem to be _____ .

A - expanding                    B - virtually unchanged           *C - shrinking

10. The upper portion of a glacier where snow is added during the winter is the _____ , at the end of the summer it is still covered with last winters snow.

*A - accumulation area          B - zone of addition              C - zone of loss

D - ablation area               E - equilibrium line

11. The lower portion of a glacier where snow is lost during the summer is the _____, at the end of the summer it is bare ice and old snow.

A - accumulation area           B - zone of addition              C - zone of loss

*D - ablation area              E - equilibrium line

12. The boundary between the portion where snow is added and where it is removed is the _____ .

A - accumulation area           B - zone of addition              C - zone of loss

D - ablation area               *E - equilibrium line

13. If the mass balance of a glacier is positive the a number of years the terminus will _____ .

*A - advance                    B - retreat                       C - remain stationary

14. Response lags are a measure of _____ .

A - climatic adjustment in a glacier       B - the size of a glacier

C - the type of glacier         *D - all of the above             E - only A and C

15. Fjord glaciers retreat by _____ .

A - overall melting    B - medial retreat    *C - calving    D - birthing    E - slumping

16. When calving begins, it progresses at a(n) _____ rate.

A - uniform             B - decelerating                 *C - accelerating

17. Temperate glaciers move primarily by _____ .

A - internal deformation of ice crystals       B - internal turbulence

*C - sliding of the basal ice over underlying rock or sediment

D - all of the above            E - only B and C

18. Internal flow of a glacier takes place mainly through _____ .

A - crystals sliding over each other       B - melting and refreezing at a lower level

*C - movement within individual ice crystals       D - gravity dispersal mechanics

E - movement as a single large crystal

19. Crevasses form in the surfacial portion of a glacier as it goes over an abrupt change in slope because _____ .

A - the surface is totally plastic          *B - the surface portion acts brittley

C - the ice is constantly recrystallizing at the surface

D - the entire glacier tears to the bedrock          E - both A and C

20. Polar glaciers move primarily by _____ .

*A - internal deformation of ice crystals          B - internal turbulence

C - sliding of the basal ice over underlying rock or sediment

D - all of the above          E - only B and C

21. Flow velocities of glaciers are on the order of _____ .

A - a few millimeters to a few centimeters per day

B - a few nannometers to a few millimeters per day

C - a few decimeters to a few meters per day

*D - a few centimeters to a few meters per day

E - a few meters to a few kilometers per day

22. An unusual episode of rapid movement or size and form change in a glacier a _____ .

A -phaser          *B - surge          C - thrust          D - push          E - rollover

23. The hypothesis postulated in your text for glacial surging is _____ .

A - the rock underlying the glacier heats and melts the glacier

B - gravity suddenly increases causing the rapid movement

*C - hydrostatic pressure at the base of the causes it to "hydroplane"

D - the basal ice crumbles allowing it to roll down slope

E - internal cleavage resistance suddenly decreases

24. As a glacier moves it acts as a _____ .

A - plow          B - file          C - sled          *D - all of the above          E - none of the above

25. Glacial striations are _____ to the direction of flow of the glacier.

*A - parallel          B - perpendicular

26. A bowl-shaped feature on a mountainside with a steep cliff on the backside and a threshold opening toward the valley

*A - cirque  B - tarn  C - arête  D - horn  E - fjord

27. The small lake that occupies a cirque.

 A - cirque  *B - tarn  C - arête  D - horn  E - fjord

28. A sharp-crested ridge formed by headward erosion of 2 glaciers.

 A - cirque  B - tarn  *C - arête  D - horn  E - fjord

29. A high, sharp-pointed peak formed by headward erosion of 3 or more glaciers.

 A - cirque  B - tarn  C - arête  *D - horn  E - fjord

30. A deep glaciated valley is flooded by the ocean at its lower end, often has a shallow sill at the seaward end.

 A - cirque  B - tarn  C - arête  D - horn  *E - fjord

31. Streamlined hills of glacially deposited material within the limits of a formed ice cap.

 A - esker  *B - drumlin  C - drift  D - till  E - erratic

32. Sediments deposited directly by a glacier is neither _____ nor _____ .

 A - graded/uniform   B - stratified/graded   *C - sorted/stratified

 D - organized/sorted   E - uniform/organized

33. The body of rock debris carried along by the sides of a glacier is the _____ .

 A - medial moraine   B - basal moraine   *C - lateral moraine

 D - ground moraine   E - edge moraine

34. The body of rock debris carried along in the middle of a glacier is the _____ .

 *A - medial moraine   B - basal moraine   C - lateral moraine

 D - ground moraine   E - end moraine

35. Nonsorted drift deposited directly from ice, random mixture of sizes, particles often are aligned with long axis parallel to the direction of flow of the ice.

 A - erratic  *B - till  C - lateral moraine   D - ground moraine

 E - end moraine

36. A glacially deposited rock that is different from the underlying bedrock.

 *A - erratic  B - till  C - lateral moraine   D - ground moraine

 E - end moraine

37. Prominent outwash terraces can usually be traced upstream to a _____ .

 A - lateral moraine   B - end moraine   C - limit of a former glacier

D - all of the above          *E - only B and C

38. Meltwater streams confined by valley walls build an outwash body called a _____ .

    A - confined moraine          B - valley moraine          *C - valley train

    D - outwash train          E - outwash moraine

39. When a glacial stream that is underloaded with sediment excavates a deeper channel through outwash deposits _____ are produced.

    A - excavation escarpments   B - outwash channels          C - excavation terraces

    *D - outwash terraces          E - outwash escarpments

40. Extremely uneven terrain underlain by ice-contact stratified drift and marked by numerous kettles and kames is clear evidence of former _____ ice conditions.

    A - massive    *B - stagnant   C- rapidly moving     D - unusually thin

    E - sediment loaded

41. In 1837, the Swiss scientist _____ proposed the hypothesis that there had been an Ice Age.

    A - John Croll          B - Milutin Milankovitch          *C - Louis Agassiz

    D - Alfred Wegener          E - Abraham G. Werner

42. During the Cenozoic Era the global climates have _____ .

    A - gradually warmed          B - remained static          *C - gradually cooled

43. As glaciers advanced toward the south the _____ rivers were forced into their present courses.

    A - Ohio and Mississippi    *B - Missouri and Ohio          C - Arkansas and Missouri

    D - Red and Arkansas          E - Mississippi and Red

44. The moisture needed to sustain the large glaciers on the continents was derived primarily from _____ .

    A - river and lakes          B - groundwater          C - the atmosphere

    D - outgassing of the mantle          *E - the oceans

45. During the most recent glacial age, world sea level fell at least _____ meters.

    A - 25          B - 50          C - 75          *D - 100          E - 125

46. A layer of glacial ice 3 Kilometers thick will cause the crust of subside approximately _____ kilometers.

    A - 2          B - 1.5          *C - 1          D - 0.75          E - 0.5

47. Classically geologists have assumed that there were _____ major glacial ages.

A - 7          B - 6          C - 5          *D - 4          E - 3

48. Recent evidence from the study of deep-sea cores suggests that there were at least _____ major glacial ages.

     A - 22          *B - 20          C - 18          D - 16          E - 14

49. Paleomagnetic evidence suggests that glacial-interglacial cycles average approximately _____ years.

     A - 60,000     B - 80,000     *C - 100,000     D - 120,000     E - 140,000

50. Glaciers tend to form and persist _____ .

     A- at high altitudes     B - at high latitudes          C - on the east side of continents

     D - all of the above     *E - only A and B

51. In the early 20th Century, _____ developed the idea that small variations in astronomical variations in Earth-Sun relationships cause temperature variations.

     A - John Croll and Abraham G. Werner          *B - Milutin Milankovitch and John Croll

     C - Louis Agassiz and Milutin Milankovitch

     D - Alfred Wegener and Louis Agassiz          E - Abraham G. Werner and Alfred Wegener

52. Analysis of gases from air bubbles trapped in modern glaciers suggests that during the Ice Ages there was _____ in the atmosphere.

     A - less oxygen and more carbon dioxide          B - less carbon dioxide and more argon

     *C - less carbon dioxide and less methane          D - less methane and less argon

     E - less methane and more oxygen

53. There appears to have been much more _____ in the atmosphere during the Ice Ages that would cause increased reflection of solar radiation back into space.

     A - water vapor          *B - dust          C - pollen          D - carbon dioxide          E - methane

## TRUE-FALSE

1. A temperate glacier has ice and water exist together in equilibrium.

     *T/F

2. If the mass balance of a glacier is positive the a number of years the terminus will advance.

     *T/F

3. Climate directly controls surges in a glacier.

     T/*F

4. Fjords are dominantly on the west side of continents.

     *T/F

270

5. Glacial striations and grooves demonstrate that the bottom of the glacier was not frozen to the bottom.

    *T/F

6. Rock flour is produced by crushing and grinding by the glacier.

    *T/F

7. The most complete record of glacial-interglacial events is on the continents.

    T/*F

8. Past surface temperatures of the Earth can be determined utilizing $^{18}O$ to $^{16}O$ isotope ratios in calcareous ooze in the deep sea.

    *T/F

9. The presence of large areas of snow and ice would increase the amount of solar radiation reflected into space and enhance the cooling rate during a glacial age.

    *T/F

## COMPLETION

1. A permanent body of ice, consisting largely of recrystallized snow, that shows evidence of downslope or outward movement due to the pull of gravity is a **glacier**.

2. Ice **temperature** is very important in controlling the way a glacier moves and its rate of movement.

3. Erratics often spreadout in a **fanlike** distribution from the source area.

## DISCUSSION

1. Describe ice as a metamorphic rock.

2. Describe internal flow in a glacier.

3. Contrast movement of temperate and polar glaciers.

4. Describe the internal flow pattern and time lapse of an ice crystal from the time it falls at the head of the glacier until it melts at the toe.

5. Discuss evidence to show that a mountain range was almost buried by an ice cap, but the tops of the mountains were not completely buried.

6. Discuss the several origins of end moraines.

7. Discuss some of the effects that lowering of sea level 100 meters during the glacial age had on worldwide distribution of land-water relationships.

8. When have pre-Pleistocene glaciations occurred?

9. Discuss the 4 geographic changes that have occurred during the history of the Earth to precipitate the onset of glacial ages.

10. Discuss the astronomical hypotheses developed by Milutin Milankovitch and John Croll in the early part of the 20th century for the initiation of glacial ages.

# CHAPTER 12

## WIND ACTION AND DESERTS

### MULTIPLE CHOICE

1. Because the Earth's surface is unevenly heated by having more heat at the equator than in polar regions the atmosphere circulates in _____ .

    A - laminar flow      *B - convection currents      C - vortices

    D - unpredictable manner      E - as a smoothly circulating envelope

2. Heated air near the equator expands, _____ .

    A - becomes lighter and sinks      B - becomes heavier and sinks

    *C - becomes lighter and rises      D - becomes heavier and rises

    E - and stays where it is

3. In the upper atmosphere air cools, _____ .

    A - becomes lighter and sinks      *B - becomes heavier and sinks

    C - becomes lighter and rises      D - becomes heavier and rises

    E - and stays where it is

4. Near the equator the air principally _____ .

    *A - rises      B - northeast      C - descends      D - west      E - east

5. From near the equator to 30° the winds are principally _____ .

    A - rises      *B - northeast      C - descends      D - west      E - east

6. From approximately 20° to 30° the upper air _____ .

    A - rises      B - northeast      *C - descends      D - west      E - east

7. Between 30° and 60° the winds are principally from the _____ .

    A - rises      B - northeast      C - descends      *D - west      E - east

8. Near the poles the air _____ .

    A - rises      B - northeast      *C - descends      D - west      E - east

9. Around the pole the winds blow from the _____ .

    A - rises      B - northeast      C - descends      D - west      *E - east

10. The polar easterlies and the westerlies meet at the _____ .

    A - trade winds       *B - polar front       C - Hadley Cell       D - Doldrums

    E - Coriolis effect

11. From near 30° to near the equator are the _____ .

    *A - trade winds       B - polar front       C - Hadley Cell       D - Doldrums

    E - Coriolis effect

12. The initial movement of sand by wind is called _____ .

    A - saltation       B - surface rolling       *C - surface creep

    D - soil creep       E - suspension

13. The process of lifting sand grains a short distance into the air and bouncing them along the ground is _____ .

    A - ground bouncing       *B - saltation       C - turbulent lifting

    D - ground turbulence       E - surfacing

14. At least 75% of the sand transported by wind in dune areas is by _____ .

    A - ground bouncing       *B - saltation       C - turbulent lifting

    D - ground turbulence       E - surfacing

15. Conditions favoring the formation of sand ripples are _____ .

    A - well-sorted sand       B - gentle wind       C - strong winds

    *D - A and B       E - A and C

16. Dust is difficult to erode because _____ .

    A - particles are so small       B - closely packed       C - smooth surface

    *D - all of the above       E - none of the above

17. Dust deposits are thicker on the _____ side of obstacles.

    A - windward       *B - leeward

18. The process of wind picking up and removing loose rock fragments, sand and dust is _____ .

    A - abrasion  B - winnowing       C - inflation  *D - deflation       E - oblation

19. When rock is impacted by wind-driven grains of sediment.

    *A - abrasion  B - winnowing       C - inflation  D - deflation       E - oblation

20. Typically, deflation rates are on the order of _____ per 1,000 years.

A - a few millimeters          *B - a few centimeters          C - a few decimeters

D - several decimeters          E - a few meters

21. The nearly continuous covering of loose stones produced by wind deflation of alluvium in a desert.

A - lag gravel          B - remnant surface          C - remnant pavement

*D - desert pavement          E - desert lag

22. Any bedrock surface or stone that has been abraded and shaped by wind-blown sediment.

A - artifact     B - verifact          *C - ventifact          D - decifact     E - centifact

23. The slope of the windward face of a dune is _____ the slope of the slip face.

A - greater than          *B - less than          C - equal to

24. Dune type is controlled by _____ .

A - the amount of sand available          B - the variability of wind direction

C - the amount of vegetation cover     *D - all of the above     E - only A and C

25. Where sand is limited and lack of moisture inhibits growth of vegetation, strong winds from one direction produce _____ dunes.

*          A - barchan     B - transverse          C - linear     D - parabolic          E - star

26. Where sand supply is greater, sinuous crested dunes oriented perpendicular to the strongest wind direction.

A - barchan     *B - transverse          C - linear     D - parabolic          E - star

27. Long, relatively straight dunes that form in areas of limited sand supply and variable wind direction.

A - barchan     B - transverse          *C - linear     D - parabolic          E - star

28. Large pyramidal dunes with sinuous radiating dunes formed by wind blowing from all directions.

A - barchan     B - transverse          C - linear     D - parabolic          *E - star

29. A dune that is U or V shaped, commonly formed in coastal dune fields, with 2 vegetation covered arms that point into the wind.

A - barchan     B - transverse          C - linear     *D - parabolic          E - star

30. Huge dune complexes that form seemingly endless and monotonous landscapes.

A - dunescape          B - ocean of sand          C - dune sea          D - migration complex

*E - sand sea

31. Thick and aerially extensive wind-deposited dust consisting largely of silt but commonly accompanied by some fine sand and clay.

  A - residuum   B - till   C - paleosol   *D - loess  E - depolith

32. Loess has the interesting property that it will _____ for long periods of time.

  A - remain airborne   *B - stand in nearly vertical cliffs

  C - fill only the bottoms of depressions  D - contain excellent marine fossils

  E - show excellent stratification

33. Loess probably derived from _____ .

  A - great desert areas that were deflated  B - deflation of recently exposed ocean floor

  C - large braided streams issuing from glaciers  D - all of the above  *E - A and C

34. Tephra dust can be distinguished from loess by _____ .

  A - grain size   B - mineralogy   C - tiny fragments of volcanic glass

  D - all of the above   *E - only B and C

35. In current usage, a desert is a region _____ .

  A - that receives less than 250 millimeters of precipitation/year

  B - evaporation exceed precipitation  C - that is arid

  *D - all of the above   E - only A and C

36. The Sahara and Great Australian Desert occur in _____ .

  *A - zone of descending air masses between 20° and 30°  B - continental interiors

  C - lee side of mountain range  D - coastal, where upwelling cold currents occur

  E - polar deserts

37. The Gobi and Takla Makan deserts are examples of _____ .

  A - zone of descending air masses between 20° and 30°  *B - continental interiors

  C - lee side of mountain range  D - coastal, where upwelling cold currents occur

  E - polar deserts

38. Most of the state of Nevada is an example of _____ .

  A - zone of descending air masses between 20° and 30°  B - continental interiors

  *C - lee side of mountain range  D - coastal, where upwelling cold currents occur

  E - polar deserts

39. Atacama Desert of Chile and Peru is an example of _____ ..

A - zone of descending air masses between 20° and 30°     B - continental interiors

C - lee side of mountain range     *D - coastal, where upwelling cold currents occur

E - polar deserts

40. Greenland and Antarctica are examples of _____ .

A - zone of descending air masses between 20° and 30°     B - continental interiors

C - lee side of mountain range     D - coastal, where upwelling cold currents occur

*E - polar deserts

41. Regolith in a desert is _____ than regolith in humid climates.

A - thinner     B - less continuous     C - coarser in texture     *D - all of the above

E - none of the above

42. Chemical weathering is less effective in a desert than a humid climate because _____ .

A - mechanical weathering is less effective     B - temperatures are generally lower

*C - there is less soil moisture     D - all of the above     E - only A and B

43. Slopes are steeper and more angular in a desert than in a humid climate because _____ .

A - average particle size is coarser     B - angle of repose is steeper

C - mechanical weathering breaks rocks along joints

*D - all of the above     E - only A and C

44. The thin, dark colored, shiny coating that accumulates on rocks in a desert is _____ .

*A - desert varnish     B - desert painting     C - desert patina

D - desert pavement     E - desert oxidation

45. The most important erosive agent in a desert is _____ .

A - wind     *B - running water     C - they are nearly equal

46. A sudden desert rainstorm is likely to produce _____ .

A - braided streams     B - gentle runoff     *C - flash floods

D - significant infiltration     E - none of the above

47. Alluvial fans form only _____ .

A - in deserts     B - at the mouth of a canyon

C - from alluvium and debris flow material     D - all of the above

*E - only B and C

48. A temporary lake on a desert valley floor is a _____ .

    A - bajada    *B - playa    C - butte    D - mesa    E - beach

49. A broad, relatively flat surface, eroded across bedrock and thinly or discontinuously veneered with alluvium, that slopes away from the base of a highland is a _____ .

    A - piedmont    B - alluvial fan    C - promontory    *D - pediment

    E - bajada

50. Both alluvial fans and pediments are _____ .

    A - convex upward    *B - concave upward    C - uniform slopes

51. Inselbergs are most likely to form _____ .

    A - relatively homogeneous rock    B - resistant rock surrounded by less resistant rock

    C - semiarid grasslands in the middle of tectonically stable continents

    *D - all of the above    E - only A and B

52. Inselbergs appear to last _____ .

    A - much shorter periods of time than the rest of the topography

    B - about the same length of time as the rest of the topography

    *C - much longer periods of time than the rest of the topography

53. Much recent desertification has been caused principally by _____ .

    A - global climatic changes    *B - human activity    C - major tectonic activity

    D - all of the above    E - only A and C

## TRUE-FALSE

1. The basic reason the atmosphere is always in motion is that more of the Sun's heat is received per unit of land surface near the equator than near the poles.

    *T/F

2. In most regions winds rarely exceed 50 km/hour.

    *T/F

3. The dustiest places on the Earth coincide with the major deserts.

    *T/F

4. Wind can erode dust is easier than sand.

    T/*F

5. Low vegetation is more efficient at trapping dust than forests.

T/*F

6. Of greatest economic importance is the deflation of bare plowed fields in farmland.

*T/F

7. Loess is an interesting unconsolidated material that will stand in nearly vertical cliffs for long periods of time.

*T/F

8. Dust from the continents is an important constituent of deep sea sediments.

*T/F

9. Deserts are randomly distributed over the face of this planet.

T/*F

10. The high temperatures of deserts cause high evaporation rates.

*T/F

11. One of the most distinctive landforms in a desert is a butte.

*T/F

12. Deserts are mostly covered by sand.

T/*F

13. A series of coalesced alluvial fans is a bajada.

*T/F

14. It is well established that pediments are formed solely by running water.

T/*F

15. Desertification is irreversible.

T/*F

## COMPLETION

1. The rate of evaporation in the deserts of southwestern United States is **10 to 20** times more than the annual precipitation.

## Discussion

1. Discuss the factors that control weather and climate.

2. If there is a dead air space 1 mm thick at ground level, how is it possible for wind to erode dust?

3. Discuss the circumstances that cause deposition of dust.

4. Describe the formation, structure, and manner of movement of dunes.

5. Discuss how to control the migration of dunes.

6. What are the 2 characteristics that demonstrate that loess is wind laid rather than water laid.

7. Discuss why land surfaces are smooth and rounded in a humid climate vs. being angular and sharp in a desert.

8. Describe the origin and structure of a playa.

9. Describe the formation of a pediment and its relation to the mountains it was formed from.

10. Describe the major symptoms of desertification.

# CHAPTER 13

## THE OCEAN MARGINS

### MULTIPLE CHOICE

1. Ocean waves and currents derive their energy from _____ .

    A - winds      B - tides      C - the sun    *D - all of the above    E - only B and C

2. Ocean currents are many miles wide by _____ meters deep.

    A - 20 to 40      B - 25 to 75      C - 60 to 80    *D - 50 to 100  E - 75 to 125

3. Ocean currents are caused when _____ .

    A - oceanic gyres are constricted    *B - the wind blows over a stretch of ocean surface

    C - the force of gravity exceeds the water strength

    D - cold and ward water masses converge       E - none of the above

4. Tides are caused by _____ .

    A - the earth's rotation      B - plate tectonics    C - drag by constant uniform winds

    *D - gravitational attraction by the moon and sun

    E - constriction of the oceans in the southern hemisphere

5. The gravitational forces from the sun causing tidal bulging are approximately _____ as effective as the moon.

    A - 1/4      *B - 1/2      C - equally    D - 2 times    E - 4 times

6. A wall of water(1+ meters high) moving rapidly up an estuary or oceanic portion of a stream is a _____ .

    A - tide front wall    B - lead edge surface   C - tidal margin surface

    *D - tidal bore      E - gravity bore

7. The size of waves depends on _____ .

    A - how fast the wind blows   B - how far the wind blows    C - how long the wind blows

    *D - all of the above      E - only A and B

8. When water is in wave motion the individual water molecules at the surface move _____ .

    A - in a linear fashion            B - in a loop twice the diameter of the wave height

C - in a loop half the diameter of the wave height

*D - in a loop equal in diameter to the wave height        E - none of the above

9. Wave base, the lower limit of bottom erosion by waves is _____ .

    A - controlled by slope of the continental shelf

    *B - equal in depth to one half of the wavelength        C - roughness of the bottom

    D - always less than 50 meters            E - temperature related

10. The geologic work of waves is mainly accomplished by _____ .

    *A - the direct action of surf   B - bottom oscillation        C - secondary rip currents

    D - refractive forces        E - breaking

11. As a wave approaches a coast the bottom is not encountered simultaneously along its length, as a result _____ .

    *A - that part of the wave slows down        B - wavelength begins to increase

    C - wave height begins to decrease    D - all of the above    E - only B and C

12. The process of a wave trend changing direction to realign parallel to the coast is _____ .

    A - wave defraction        *B - wave refraction        C - wave infraction

    D - wave subtraction        E - wave extraction

13. Along a shoreline the forces of _____ and _____ are in opposition to each other.

    A - gravity/deposition of sediment        *B - deposition of sediment/erosion

    C - erosion/wave refraction        D - wave refraction/ mass wasting

    E - mass wasting/gravity

14. Wave erosion takes place _____ .

    A - at sea level in the surf zone        B - below sea level        C - above sea level

    *D - all of the above            E - only B and C

15. The maximum depth limit for vigorous erosion by surf is _____ meters.

    A - 3        B - 6        *C - 9        D - 12        E - 15

16. Even though storm waves are capable of exerting incredible energy, most coastal erosion is confined to a zone _____ meters above and below mean sea level.

    A - 3        B - 6        C - 8        *D - 10        E - 14

17. Blocks of rock are dislodged from sea cliffs by _____ .

    A - the force of the water breaking the rock

B - waves picking up blocks and dashing them into the sea cliff

*C - waves trapping air in fissures, compressing it, forcing blocks

D - all of the above          E - none of the above

18. Waves strike shoreline at some oblique angle causing _____ .

A - transverse currents and rip currents          B - rip currents and surf

*C - surf and longshore currents          D - longshore currents and tidal bores

E - tidal bores and transverse currents

19. The zig-zag movement of sediment up and down a beach caused by waves striking the beach at an oblique angle _____ .

A - beach placer          *B - beach drift          C - longshore drift

D - beach displacement          E - beach dissipation

20. Beach placers of heavy minerals are concentrated by _____ .

A - beach drift          B - longshore currents          C - surf

D - both A and B          *E - both B and C

21. Seaward of the surf zone, in deeper water, bottom sediment is moved by _____ .

A - currents          B - mass wasting          C - unusually large storm waves

D - only A and B          *E - only A and C

22. As sediment is moved into deeper water, a given size is picked up less often because there is less energy, as a result _____ .

A - sediment is sorted getting finer in a seaward direction

B - ledges form where the water runs out of energy

C - sediment is size sorted          D - all of the above          *E - only A and C

23. On a low, open sandy beach the zone between lowest tide and average high-tide is the ____ .

A - surf zone   *B -foreshore          C - beach deposit          D - berm          E - backshore

24. On a low, open sandy beach the zone of wave deposition that is nearly horizontal or sloping landward is the _____ .

A - surf zone   B -foreshore          C - beach deposit          *D - berm          E - backshore

25. The zone behind the horizontal beach crest inland as far as the surf zone extends is the _____ .

A - surf zone   B -foreshore          C - beach deposit          D - berm          *E - backshore

26. Wave-cut cliffs form by _____ .

A - abrasion of rock surfaces by sediment loaded surf

B - large rocks being thrown against cliffs by storm waves

*C - cliffs being undercut and the upper portion collapsing

D - rapid lithification of sediment that has been piled up

E - erosion by longshore currents

27. If sediment is deposited faster than it can be eroded the coastline will _____ .

   *A - prograde      B - agrade      C - degrade      D - deltagrade      E - subgrade

28. An elongated ridge of sand or gravel that projects from land(often a beach) and ends in open water.

   A - stack      *B - spit      C - tombolo      D - bay barrier      E - hook

29. A depositional ridge that is built from an island to the mainland or another island.

   A - stack      B - spit      *C - tombolo      D - bay barrier      E - hook

30. A depositional ridge that is built is built across the mouth of a bay where the currents are too weak to erode it.

   A - stack      B - spit      C - tombolo      *D - bay barrier      E - hook

31. Old berms that were constructed during storms along sandy coasts.

   A - stack      B - spit      C - tombolo      D - bay barrier      *E - beach ridges

32. A body of sand that is parallel to a sea coast and has been piled up above sea-level by waves breaking on a sandy bottom and modifying the longshore bars.

   A - stack      *B - barrier island      C - tombolo      D - bay barrier

   E - beach ridges

33. The water body behind a barrier island or other obstruction.

   *A - lagoon      B - barrier island      C - tombolo      D - bay barrier

   E - beach ridges

34. Organic reefs are built _____ .

   A - at or close to sea level           B - low latitudes, 18° C or more water

   C - clear water           *D - all of the above           E - only A and C

35. A reef either attached to or closely bordering the adjacent land.

   A - barrier reef           *B - fringing reef           C - atoll

36. A reef separated from the land by a lagoon of considerable size.

   *A - barrier reef           B - fringing reef           C - atoll

284

37. A roughly circular tropical island built on top of a volcano.

    A - barrier reef               B - fringing reef             *C - atoll

38. Sea level changes on long term basis because of _____ .

    A - changes in the volume of water in ocean basins because of glaciers

    B - volume changes in the ocean basins because of lithospheric plates

    C - changes in the total amount of water on the surface of the Earth

    D - all of the above        *E - only A and B

39. The large number of shoreline feature found on the sea floor today is indicative of _____ of the shoreline.

    *A - submergence     B - emergence     C - divergence     D - convergence

    E - colovergence

40. Shoreline features that stand higher than present shorelines is indicative of _____ of the shoreline.

    A - submergence    *B - emergence     C - divergence     D - convergence

    E - colovergence

41. Tsunami may be caused by _____ .

    A - earthquakes near or under water      B - submarine landslides

    C - submarine volcanic eruptions    *D - all of the above     E - none of the above

42. Cliffed shorelines are susceptible to frequent landslides because _____ .

    A - earthquakes are becoming more abundant

    B - the activity of all the people tends to dislodge the cliffs

    *C - erosion eats the base of the cliffs away allowing mass wasting

    D - all of the above      E - only A and B

43. Landslides caused by collapse of large sea cliffs may cause _____ .

    A - large magnitude earthquakes      B - burial of large populated areas

    *C - very large waves that do serious local damage

    D - all of the above      E - only A and B

44. Seacliffs can be protected by _____ .

    A - covering the cliff with large boulders

    B - constructing seawalls with deep footings parallel to the cliff

C - breakwaters          D - groins          *E - only A and B

45. Entire beach systems can be very effectively and economically protected by the construction of _____ .

A - breakwaters      B - groins      C - seawalls      D - all of the above

*E - modification at one place causes disaster somewhere else

## TRUE-FALSE

1. Shorelines are quite static.

    T/*F

2. There are 2 tidal bugles on opposite sides of the Earth.

    *T/F

3. Breaking waves all look alike.

    T/*F

4. The general tendency is for an irregular coast to become smoother and less indented.

    *T/F

5. Surf is effectively like a knife edge or saw cutting horizontally into the land.

    *T/F

6. Longshore currents have a constant direction all year.

    T/*F

7. The greater the angle of waves to shore, the greater the amount of longshore drift.

    *T/F

8. The Mississippi Delta is a very simple structure.

    T/*F

9. Beach ridges are all more than 5,000 years old.

    T/*F

10. Following melting of the glaciers during the last ice age sea level rose slowly enough that all former shoreline features were destroyed..

    T/*F

11. Unraveling the history of sea-level fluctuations along a coast can be a difficult and challenging exercise.

    *T/F

12. Beaches are very delicately balanced and any sort of modification in the area may spell disaster to a beach.

     \*T/F

# DISCUSSION

1. Describe the major oceanic circulation pattern and the types of water bodies involved.

2. Describe how a wave breaks.

3. Discuss erosional activity in the surf zone.

4. Discuss the economics of beach placers, type of mineral and site of placer.

5. Describe the formation of a wave-cut cliff and wave-cut beach.

6. Discuss the 3 conditions that cause beaches to be very steep.

7. Explain why the Mississippi River has a large delta, but the Columbia River has no delta.

8. What is the probable maximum absolute age and the relationship of beach ridges in western Alaska to invasion of ancient peoples.

9. Describe that evolution of an atoll.

10. What factor control the configuration of a coastline?

11. Describe the characteristics of a tsunami that make them such hazards to life.

12. Discuss why beach protection/modification is so expensive and fruitless.

# CHAPTER 14

## DEFORMATION OF ROCKS

### MULTIPLE CHOICE

1. Equal pressure in all directions is _____ .

　　　*A - uniform stress　　　　B - confining stress　　　　C - differential stress

　　　D - tensional stress　　　　E - compressional stress

2. Uniform pressure inside a massive rock body _____ .

　　　A - uniform stress　　　　*B - confining stress　　　　C - differential stress

　　　D - tensional stress　　　　E - compressional stress

3. Pressure that is not uniform in all directions _____ .

　　　A - uniform stress　　　　B - confining stress　　　　*C - differential stress

　　　D - tensional stress　　　　E - compressional stress

4. Pressure that stretches rocks _____ .

　　　A - uniform stress　　　　B - confining stress　　　　C - differential stress

　　　*D - tensional stress　　　　E - compressional stress

5. Pressure that squeezes rocks _____ .

　　　A - uniform stress　　　　B - confining stress　　　　C - differential stress

　　　D - tensional stress　　　　*E - compressional stress

6. Pressure that causes slippage and translation _____ .

　　　*A - shear stress　　　　B - confining stress　　　　C - differential stress

　　　D - tensional stress　　　　E - compressional stress

7. The term used to describe the deformation of a rock is _____ .

　　　A - stress　　　*B - strain　　　C - elastic limit　　　D - ductile deformation

　　　E - fracture

8. The term used to describe pressure on a rock is _____ .

　　　*A - stress　　　B - strain　　　C - elastic limit　　　D - ductile deformation

E - fracture

9. When a solid changes size, but not shape is _____ .

    *A - uniform stress           B - confining stress          C - differential stress

    D - tensional stress          E - compressional stress

10. When a solid changes shape and may or may not change size is _____ .

    A - uniform stress           B - confining stress        *C - differential stress

    D - tensional stress          E - compressional stress

11. Reversible, or nonpermanent, change in volume and/or shape of a stressed rock is _____ .

    A - stress               *B - elastic deformation      C - elastic limit

    D - ductile deformation      E - fracture

12. The point beyond which a stressed rock will not return to its original shape or size _____ .

    A - stress      B - strain      *C - elastic limit      D - ductile deformation

    E - fracture

13. An irreversible change in shape and/or volume of a rock that has been stressed beyond the elastic limit is _____ .

    A - stress      B - strain      C - elastic limit      *D - ductile deformation

    E - fracture

14. When the limits of both elastic and ductile deformation are exceeded in a solid is _____ .

    A - stress      B - strain      C - elastic limit      D - ductile deformation

    *E - fracture

15. A brittle substance tends to deform by _____ .

    A - change of shape    *B - fracture    C - stress      D - composition

    E - temperature

16. A ductile substance tends to deform by _____ .

    *A - change of shape    B - fracture    C - stress      D - composition

    E - temperature

17. The higher temperature, the more _____ and less _____ a solid becomes.

    A - elastic/ ductile       *B - ductile/ brittle      C - brittle/ flexible

    D - flexible/ plastic      E - plastic/ elastic

18. Near the surface rocks are _____, but at depth, where temperatures are high because of the geothermal gradient, rocks become _____ .

A - ductile/ elastic      *B - brittle/ ductile      C - flexible/ brittle

D - plastic/ flexible      E - elastic/ plastic

19. High confining stress hinders the formation of _____ and so reduces brittle properties.

A - change of shape      *B - fracture      C - stress

D - composition      E - temperature

20. If the stress builds up slowly and gradually and is maintained for a long period, the atoms have time to move, and the solid can slowly readjust and change shape by _____ .

A - stress      B - strain      C - elastic limit

*D - ductile deformation      E - fracture

21. The rate at which a rock is forced to change shape or volume is _____ .

A - stress      *B - strain rate      C - elastic limit

D - ductile deformation      E - fracture

22. Rocks that tend to deform by ductile deformation are _____ .

A - limestone, marble      B - sandstone, quartzite      C - shale, slate

D - all of the above      *E - only A and C

23. Rocks that tend to deform by brittle deformation are _____ .

A - granite, granodiorite      B - sandstone, quartzite      C - shale, slate

D - all of the above      *E - only A and B

24. A fracture in a rock along which movement occurs is a _____ .

A - joint      *B - fault      C - fold      D - shear      E - strike

25. A fracture in a rock along which no movement occurs is a _____ .

*A - joint      B - fault      C - fold      D - shear      E - strike

26. Typically movement along a fault occurs _____ at a time.

*A - a few centimeters to a few meters      B - a few meters to a few tens of meters

C - a few tens of meters to a few hundred meters

D - a few hundred meters to a few kilometers

E - a few kilometers to a few hundred kilometers

27. Abrupt fault movement may be _____ .

A - horizontal          B - vertical          C - oblique

D - all of the above one at a time          *E - all of the above at once

28. The study of rock deformation come under the subfield of geology ____ .

A - historical geology          B - sedimentology          *C - structural geology

D - stratigraphy          E - paleontology

29. The direction of a horizontal line on the surface of a planar rock feature is _____ .

A - dip          B - outcrop          C - exposure          *D - strike          E - joint

30. The angle measured from horizontal down to a planar rock feature and perpendicular to a horizontal line on the surface of the feature is _____ .

*A - dip          B - outcrop          C - exposure          D - strike          E - joint

31. When standing on a fault the block over your head is the _____ .

*A - hanging wall block          B - graben          C - rift          D - footwall block

E - horst

32. When standing on a fault the block you stand on is the _____ .

A - hanging wall block          B - graben          C - rift          *D - footwall block

E - horst

33. In a normal, or tensional, fault the hanging wall block goes ____ relative to the footwall block.

A - up          *B - down          C - laterally

34. In a reverse, or compressional, fault the hanging wall block goes _____ relative to the footwall block.

*A - up          B - down          C - laterally

35. A down-dropped block between 2 faults with parallel strikes is a _____ .

A - graben          B - rift          C - horst

D - all of the above          *E - only A and B

36. A down-dropped block between 2 faults with parallel strikes, but only one side goes down and the other side hangs-up, is a _____ .

*A - half-graben          B - rift          C - horst

D - all of the above          E - only A and B

37. An upthrust block between 2 faults with parallel strikes is a _____ .

A - graben          B - rift          *C - horst

D - all of the above          E - only A and B

38. The greatest system of active grabens on Earth today is in _____ .

   A - western North America          *B - east Africa          C - central Asia

   D - western South America          E - eastern Australia

39. A compressional fault with a dip of less than 15° is a _____ fault.

   A - graben     B - rift     C - horst     *D - thrust     E - only A and B

40. A fault that has principally horizontal movement rather than vertical movement is a _____ fault.

   A - graben     B - rift     C - horst     *D - strike-slip     E - only A and B

41. A strike-slip fault is _____ lateral, if when you look across the fault the other side has moved toward your right.

   *A - right     B - left     C - horizontal     D - controlled

   E - uncontrolled

42. A strike-slip fault is _____ lateral, if when you look across the fault the other side has moved toward your left.

   A - right     *B - left     C - horizontal     D - controlled

   E - uncontrolled

43. Strike-slip faults that form tectonic plate boundaries are called _____ faults.

   A - translation          B - teleform          *C - transform

   D - transposed          E - translated

44. A polished, grooved or striated surface on a fault plane is _____ .

   A - polishing          B - smoothing          C - siding

   D - grinding          *E - slickensides

45. The 2 sides of a fault may be broken-up into _____ .

   A - fault conglomerate     *B - fault breccia     C - disaggregation zones

   D - roughage zones          E - collapse breccia

46. The bending or warping of rocks is referred to as _____ .

   A - bending          B - warping          *C - folding

   D - wrapping          E - curving

47. A fold in rock where relatively flat-lying rocks have been draped into a sloping surface, but the rocks coming into, and going out of, the fold are parallel to each other.

*A - monocline        B - anticline  C - homocline        D - syncline   E - dispocline

48. A fold where the rocks have been up arched structure.

A - monocline        *B - anticline  C - homocline        D - syncline   E - dispocline

49. A fold where the rocks have been forced into a sag-like structure.

A - monocline        B - anticline  C - homocline        *D - syncline  E - dispocline

50. The 2 sloping sides of a fold are the _____ .

A - axis        B - plunge    *C - limbs    D - plunging fold    E - axial plane

51. A median line along the crest or trough of a fold and equally between the sloping sides _____ .

*A - axis        B - plunge    C - limbs    D - plunging fold    E - axial plane

52. A fold with an inclined axis is a _____ .

A - axis        B - plunge    C - limbs    *D - plunging fold    E - axial plane

53. The angle between a fold axis and horizontal is the _____ .

A - axis        *B - plunge    C - limbs    D - plunging fold    E - axial plane

54. An imaginary plane that divides a fold as symmetrically as possible is _____ .

A - axis        B - plunge    C - limbs    D - plunging fold    *E - axial plane

55. In a deeply eroded folded mountain system, topography is largely determined by _____ .

A - folding    B - faulting    *C - lithology  D - all of the above    E - none of the above

## TRUE-FALSE

1. When the elastic limit is exceeded the substance will not return to its original size/shape.

    *T/F

2. The higher the temperature, the more ductile and less brittle a solid becomes.

    *T/F

3. The rate at which a solid is strained is just as significant as how long a stress is active.

    *T/F

4. Failure by fracture is common in upper-crustal rocks.

    *T/F

5. Composition makes no difference in determining whether a rock deforms in a ductile or brittle manner.

    T/*F

6. Movement of thousands of meters along a fault occurs a few centimeters or meters at a time.

   *T/F

7. Geologic maps are easy to construct because all the rock is exposed.

   T/*F

8. Absolute direction of movement on a fault is easily determined.

   T/*F

9. Many of the largest and most active faults are strike-slip faults.

   *T/F

10. A compressional fault with a dip of less than 15° is called a thrust fault.

    *T/F

11. The very intense folding that can be so widely observed in mountain ranges probably occurred below the brittle-ductile transition when the rocks were deeply buried and thus were subjected to high temperature and high confining stress.

    *T/F

12. Anticlines and synclines are usually paired.

    *T/F

13. All folds are symmetrical.

    T/*F

14. It is always easy to determine whether a rock unit is right-side up or upside down.

    T/*F

15. The great oil pools of the Middle East are structural traps.

    *T/F

## DISCUSSION

1. What are the essential conditions controlling the relative importance of ductile properties vs. brittle properties?

2. Why is time important in whether a material will under go brittle deformation or ductile deformation.

3. Discuss how composition effects rock deformation properties.

4. Describe the manner in which a fault moves and why movement is not continuous.

5. Describe how strike and dip is determined and what it means.

6. Briefly describe how a geologic map is made.

7. Describe the meaning of the terms hanging wall block and footwall block.

8. How can you tell whether a break in a rock outcrop is a joint or if there has been movement along it?

9. Discuss the several relationships between folding and faulting.

10. Discuss the relationship between underlying structural features and lithology and topography.

11. Discuss the 5 essential requirements for an oil pool to form.

# CHAPTER 15

## EARTHQUAKES AND THE EARTH'S INTERIOR

### MULTIPLE CHOICE

1. Most earthquakes occur in _____ .

    A - ductile rocks of the lithosphere      *B - brittle rocks of the lithosphere

    C - ductile rocks of the asthenosphere      D - brittle rocks of the asthenosphere

    E - none of the above

2. The study of earthquakes is _____ .

    A - sideralology      B - spieleology      *C - seismology

    D - sphenology      E - none of the above

3. The device that records earthquakes is a _____ .

    A - sideralograph      B - spieleograph      *C - seismograph

    D - sphenograph      E - none of the above

4. The 2 most widely used types of seismographs are _____ .

    A - mass/volume      B - James stress/ mass      C - Benioff strain/ James stress

    *D - inertial/ Benioff strain      E - volume/ inertial

5. The place where energy is first released to cause an earthquake is called the _____ .

    A - epicenter   B - center of motion   C - site of motion    *D - focus    E - origin

6. The spot of the Earth's surface directly above where the movement causing the earthquake occurred.

    *A - epicenter   B - center of motion   C - site of motion    D - focus    E - origin

7. The fastest body earthquake waves, move by changing the volume of the rocks.

    A - compressional    B - shear     C - P-waves    D - S-waves    *E - only A and C

8. P-waves will travel through _____ .

    A - gas      B - liquid      C - solid     *D - all of the above    E - only B and C

9. S-waves will travel through _____ .

    A - gas      B - liquid     *C - solid     D - all of the above    E - only B and C

10. The slower body earthquake waves, move by changing the shape of the rocks.

A - compressional    B - shear    C - P-waves    D - S-waves    *E - only B and D

11. The phenomenon of seismic waves bouncing off of a material of different density is _____ .

A - defraction    B - refraction    C - deflection    *D - reflection    E - infraction

12. The phenomenon of seismic waves bending as they go from material of one density to a different density is _____ .

A - defraction    *B - refraction    C - deflection    D - reflection    E - infraction

13. The property of surface seismic waves traveling at ever greater speed as greater density of material is encountered and therefore taking different paths is referred to as _____ .

A - defraction    B - refraction    C - deflection    D - reflection    *E - dispersion

14. To accurately locate an earthquake epicenter, a minimum of _____ seismographs are needed.

A - 2    *B - 3    C - 4    D - 5

E - it makes no difference because earthquakes cannot be located

15. Very large earthquakes are likely to occur approximately once every _____ years at any given site.

A - 50    B - 75    *C - 100    D - 125    E - 150

16. The Richter magnitude scale measures _____ .

A - height of vibrations on the ground    B - magnitude of ground waves

C - magnitude of surface waves on the seismogram

*D - magnitude of P and S waves on the seismogram    E - none of the above

17. The Richter scale is _____ .

A - arithmetic    B - geometric    *C - logarithmic

D - trigonometric    E - spherical

18. Each unit increase in magnitude on the Richter scale corresponds to a ____fold increase in the amplitude of the wave signal.

A - 2    B - 4    C - 6    D - 8    *E - 10

19. The largest earthquakes to hit in the United States was at ____ .

A - New York City    B - Charleston, South Carolina    C - San Francisco

*D - New Madrid, Missouri    E - Los Angeles

20. Approximately 80% of all earthquakes originate in the _____ .

A - pan-Arctic belt          B - circum-Atlantic belt          *C - circum-Pacific belt

D - Mediterranean-Himalayan belt          E - midocean ridges

21. The deepest earthquakes, up to 700 km deep, are associated with _____ .

A - midocean ridges          B - transform faults          *C - deep trenches

22. Most earthquakes are brittle fracture and are less than _____ km deep.

A - 50          B - 75          *C - 100          D - 125          E - 150

23. The sloping surface below the trenches where the plate is being subducted is the _____ .

A - Richter zone          B - Jamison zone          C - Walcott zone          D - Powell zone

*E - Benioff zone

24. If the first portion of a P-wave to arrive at a seismograph is a compressive pulse, the release of energy and movement on the fault was _____ the seismograph.

*A - toward          B - away from          C - parallel          D - there is no way of knowing

25. If the first portion of a P-wave to arrive at a seismograph is a expansion, the release of energy and movement on the fault was _____ the seismograph.

A - toward          *B - away from          C - parallel          D - there is no way of knowing

26. Lithosphere under tension, normal faults, low Richter magnitudes, foci less than 100 km.

A - transform fault boundaries          *B - spreading center          C - convergent boundaries

27. Huge, vertical, strike-slip faults, foci less than 100 km, high Richter magnitudes.

*A - transform fault boundaries          B - spreading center          C - convergent boundaries

28. Seismic wave velocities are controlled by _____ of the rock.

A - density          B - porosity          C - rigidity          D - all of the above          *E - only A and C

29. Distinct discontinuities in the interior of the Earth can be identified from _____ of body waves.

A - refraction          B - reflection          C - deflection          D - all of the above          *E - only A and B

30. The boundary at the base of the crust is the _____ discontinuity.

A - Benioff          B - Kimberlite          C - Fox          *D - Mohorovicic          E - Strong

31. The thickness of the ocean crust is less than _____ km thick.

A - 3          B - 7          *C - 10          D - 14          E - 19

32. The composition of the ocean crust is _____ .

A - basalt          B - granite          C - gabbro          D - all of the above          *E - only A and C

33. The thickness of the continental crust is _____ km thick.

298

A - 10 to 40      *B - 20 to 60          C - 30 to 50      D - 40 to 60              E - 50 to 80

34. The composition of the continental crust is dominantly _____ .

   A - diorite      B - granite      C - gabbro      D - all of the above      *E - only A and B

35. The mantle appears to have a lot of the minerals like _____ .

   A - quartz and muscovite      *B - olivine and pyroxene      C - amphibole and biotite

   D - kaolinite and calcite      E - graphite and dolomite

36. At the mantle-core boundary P-waves are strongly _____ .

   A - reflected    B - absorbed    C - refracted    D - all of the above      *E - only A and C

37. At the mantle-core boundary S-waves are completely _____ .

   A - reflected    *B - absorbed    C - refracted    D - all of the above      E - only A and C

38. The outer core appears to be _____ .

   A - solid iron          B - solid basalt          C - solid nickel          *D - molten iron

   E - molten nickel

39. The inner core appears to be _____ .

   *A - solid iron          B - solid basalt          C - solid nickel          D - molten iron

   E - molten nickel

40. The low-velocity zone in the mantle extends approximately over the range _____ km.

   A - 50 to 260    B - 75 to 300    C - 90 to 325    *D - 100 to 350          E - 110 to 400

41. The low-velocity zone probably exists because _____ .

   A - the composition is different      *B - it has undergone partial melting

   C - the density decreases      D - all of the above      E - only A and C

42. The top of the low-velocity zone coincides with _____ .

   *A - the base of the lithosphere      B - the base of the mesosphere

   C - the base of the crust          D - none of the above

43. The 400 km seismic discontinuity appears to be caused by _____ .

   A - a composition change      *B - the transition from olivine to its polymorph spinel

   C - the change to a more rigid medium      D - all of the above      E - only A and C

44. The 670 km seismic discontinuity is caused by _____ .

   A - a compositional change          B - a polymorphic change

C - slabs of lithosphere floating at that depth

D - all of the above          *E - the cause is unknown

45. An area with a gravity reading different from what it should be is a _____ .

    A - gravity block          *B - gravity anomaly          C - gravity discontinuity

    D - gravity discordance          E - gravity merge

46. When the crust gets topographically higher, the base _____ .

    A - also gets higher          B - remains flat          *C - gets deeper

47. When the crust gets thicker it produces a _____ gravity anomaly.

    A - positive          B - nonexistent          *C - negative

48. The property of floatational balance among segments of the lithosphere is _____ .

    A - isometrics          B - isoclinal          C - isogonal

    *D - isostasy          E - isobouyant

49. The continental crust must always stand high and the oceanic crust must always remain low because of _____ .

    A - difference in thickness          B - difference in age          *C - difference in density

    D - all of the above          E - none of the above

## TRUE-FALSE

1. Sudden movement along faults seems to be the cause of most earthquakes.

    *T/F

2. Seismic body wave velocities are a function of the density of the medium they are passing through.

    *T/F

3. Earthquakes that cause major loss of life occur frequently.

    T/*F

4. From seismograph records it is possible to tell the direction of movement on a fault.

    *T/F

5. Only P-waves can be utilized to tell direction of movement on a fault.

    T/*F

6. Collision zones are typically several hundred km wide.

    *T/F

7. The composition of the mantle is well understood.

   T/*F

8. The upper and lower surfaces of the low-velocity zone in the mantle is quite sharp.

   T/*F

9. The Earth is an ellipsoid the is slightly flattened at the poles and bulges at the equator.

   *T/F

10. Gravimeters are very similar to an inertial seismograph.

   *T/F

## COMPLETION

1. The most widely accepted theory concerning the origin of earthquakes involves slipping faults and the **elastic rebound** theory.

2. The study of earthquakes is **seismology**.

3. The place where energy is first released to cause an earthquake is called the **earthquake focus**.

4. Each unit increase in magnitude on the Richter scale corresponds to a **ten**fold increase in the amplitude of the wave signal.

## DISCUSSION

1. Discuss the elastic rebound theory of earthquake origin.

2. Describe the principle inertial seismographs are based upon.

3. Discuss the paths of travel of seismic waves as they travel through the body of the Earth.

4. Describe the procedure to locate an earthquake epicenter.

5. Discuss the 6 principal kinds of earthquake dangers.

6. Describe the earthquake motion and activity in a convergent boundary zone.

7. How did Mohorovicic discover the discontinuity at the base of the crust?

8. Explain why the inner core is solid even though it is much hotter than the outer core.

# CHAPTER 16

## GLOBAL TECTONICS

### MULTIPLE CHOICE

1. The name of the last great supercontinent proposed by Alfred Wegener is _____ .

    A - Gondwanaland    *B - Pangaea    C - Laurasia    D - Wrangelia        E - Atlantus

2. Alfred Wegener proposed that the last great supercontinent existed during the _____ Period.

    *A - Permian            B - Pennsylvanian                C - Triassic

    D - Ordovician          E - Cretaceous

3. Alfred Wegener's great supercontinent broke apart into a northern continent called _____ .

    A - Gondwanaland    B - Pangaea    *C - Laurasia    D - Wrangelia        E - Atlantus

4. Alfred Wegener's great supercontinent broke apart into a southern continent called _____ .

    *A - Gondwanaland    B - Pangaea    C - Laurasia    D - Wrangelia        E - Atlantus

5. One of Alfred Wegener's principle lines of evidence for the existence of Pangaea was _____ .

    A - seismic history    B - volcanic activity patterns        *C - continental ice sheets

    D - distribution of dinosaur remains        E - radiometric age dates

6. Paleomagnetic studies of rocks will show _____ at the time of formation.

    A - the temperature            B - the longitude of the site of formation

    C - the approximate depth of formation

    *D - whether the Earth's magnetic field was normal or reversed

    E - absolute strength of the Earth's magnetic field

7. Paleomagnetic studies of rocks will show _____ at the time of formation.

    A - the temperature            B - the longitude of the site of formation

    C - the approximate depth of formation

    *D - the orientation of the magnetic poles relative to site of formation

    E - absolute strength of the Earth's magnetic field

8. Paleomagnetic studies of rocks will show _____ at the time of formation.

A - the temperature          B - the longitude of the site of formation

C - the approximate depth of formation     *D - the magnetic latitude

E - absolute strength of the Earth's magnetic field

9. In 1962, Harry Hess proposed the idea of _____ .

A - continental drift      B - apparent polar wandering      C - hot spots

D - mantle convection      *E - seafloor spreading

10. According to the theory of plate tectonic new oceanic crust is formed _____ .

A - at the edge of continents      *B - along the mid-ocean ridges

C - in the Benioff zones associated with deep oceanic trenches

D - on the abyssal floor      E - from the continental craton

11. According to the theory of plate tectonic old oceanic crust is destroyed _____ .

A - at the edge of continents      B - along the mid-ocean ridges

*C - in the Benioff zones associated with deep oceanic trenches

D - on the abyssal floor      E - from the continental craton

12. The most recent magnetic reversal recorded near the crest of a mid-ocean ridge occurred _____ years ago.

A - 456,000    *B - 730,000    C - 982,000    D - 1,247,000      E - 2,327,000

13. The oldest reversals found in the oceanic crust dates to _____ .

A - middle Pliocene      B - early Paleocene      C - late Cretaceous

*D - middle Jurassic      E - middle Triassic

14. A lithospheric plates with a small amount of continental crust moves at relatively _____ velocity compared to a lithospheric plate with a large amount of continental crust.

*A - high          B - equal          C - low

D - there is no way of knowing this      E - there is no pattern in the rates

15. A lithospheric plates with a large amount of continental crust moves at relatively _____ velocity compared to a lithospheric plate with a small amount of continental crust.

A - high          B - equal          *C - low

D - there is no way of knowing this      E - there is no pattern in the rates

16. Lithospheric plates in _____ motion.

A - straight line      B - unpredictable      *C - rotational

D - spherical      E - stationary

303

17. Mid-ocean ridges that are spreading at a slow rate have a central rift that is _____ than on mid-ocean ridges that are spreading at a fast rate.

   *A - wider and more pronounced          B - narrower and more pronounced

   C - narrower and less pronounced          D - wider and less pronounced

   E - they are all the same

18. A large fraction of the heat that escapes from the Earth's interior does so _____ .

   A - along the trenches          *B - along the spreading centers

   C - on the deep sea floor          D - along the continental rise

   E - in the interior of continents

19. What happens as a plate sinks into the asthenosphere at a convergent margin is inferred from _____ .

   A - seismic studies          B - deformed rocks where collision occurs

   C - type of volcanism          *D - all of the above          E - only B and C

20. Wet partial melting of the water saturated upper layer of the oceanic crust during subduction forms _____ magma.

   A - rhyolitic          *B - andesitic          C - basaltic          D - all of the above

   E - none of the above

21. The magmatic arc is parallel to the seafloor trench and separated by a distance of 100 to 400 km depending on _____ .

   A - the composition of the descending crust          B - the amount of water present

   C - the temperature of the interior          D - the thickness of the continental crust

   *E - angle of dip of the descending plate

22. Old, cold, dense oceanic crust forms island arcs that have a _____ radius of curvature.

   A - small          B - moderate          *C - large          D - variable          E - unknown

23. Young, warmer, less dense oceanic crust forms island arcs that have a _____ radius of curvature.

   *A - small          B - moderate          C - large          D - variable          E - unknown

24. The down-going slab of subducting oceanic crust remains brittle to a depth of approximately _____ km.

   A - 400          B - 500          C - 600          *D - 700          E - 800

25. Mélange are characterized by _____ schists.

A - black       B - green       C - brown       D - red       *E - blue

26. A chaotic mixture of broken, jumbled, crushed, sheared, thrust-faulted sedimentary and high-pressure, low-temperature metamorphic rock is a _____ .

   A - menagerie       *B - mélange       C - mendacious       D - mendicant

   E - meniscus

27. Metamorphism in a mélange is commonly _____ .

   A - high-pressure, high-temperature       B - low-pressure, high-temperature

   *C - high-pressure, low-temperature       D - low-pressure, low-temperature

   E - unpredictable

28. The faults at plate margins are _____ faults

   *A - transform       B - transcurrent       C - transcendent       D - transient

   E - translation

29. Transform faults are _____ .

   A - huge       B - vertical       C - strike-slip       *D - all of the above       E - A and B

30. Transform faults are characterized by _____ .

   A - large numbers of shallow-focus earthquakes       B - abundant volcanism

   C - zones of shattered rock       D - all of the above       *E - only A and C

31. The best known transform fault that separates the North American and the Pacific plates is the _____ .

   A - St. Ann's       B - San Francisco       C - Santa Inez       D - Santo Domingo

   *E - San Andreas

32. A study of hot spots suggests that the _____ Plate is nearly stationary.

   A - Pacific       B - Australian-Indian       C - Asian       *D - African

   E - South American

33. The _____ Plate appears to be getting smaller.

   *A - Pacific       B - Australian-Indian       C - Asian       D - African

   E - South American

34. The best reference available for determining absolute plate motions is/are _____ .

   A - transform faults       B - mid-ocean ridges       *C - hot spots

   D - deep-sea trenches       E - nothing really seems to work

35. The core of very ancient rock in a continent is the _____ .

    A - orogen     B - continental shield     *C - craton     D - stable platform

    E - décollement

36. Elongate regions of crust that have been intensely folded and faulted during continental collisions are _____ .

    *A - orogen     B - continental shield     C - craton     D - stable platform

    E - décollement

37. The best evidence available to support the idea that plate tectonics operated at least as far back as 2 billion years ago is _____ .

    A - the distribution of fossils     *B - the existence of ancient collision belts

    C - unaltered sedimentary rocks     D - all of the above

    E - no such evidence exists

38. Several ancient cratons and orogens have been identified in the Canadian Shield through _____ .

    A - detailed fossil studies     B - careful mapping     C - radiometric age dating

    D - only A and C     *E - only B and C

39. A _____ margin occurs in the stable interior of a plate far from the plate margins.

    *A - passive continental     B - continental convergent     C - continental collision

    D - transform fault continental     E - accreted terrane continental

40. A typical depositional sequence(from oldest to youngest) along a passive continental margin would be _____ .

    A - evaporites - nonmarine clastics - marine clastics

    B - nonmarine clastics - marine clastics - evaporites

    C - marine clastics - evaporites - nonmarine clastics

    *D - nonmarine clastics - evaporites - marine clastics

    E - evaporites - marine clastics - nonmarine clastics

41. A _____ margin with intense deformation of continental margin with stratovolcanos and sediments in mélange.

    A - passive continental     *B - continental convergent     C - continental collision

    D - transform fault continental     E - accreted terrane continental

42. Great mountain systems that lie in the interior of a major landmass, formed from marine sediments that have been intensely folded and thrust faulted.

A - passive continental     B - continental convergent   *C - continental collision

D - transform fault continental     E - accreted terrane continental

43. The collision line between the two masses of deformed sediment caught up in a continental collision is commonly marked by the presence of _____ .

*A - serpentinites     B - carbonatites     C - amphibolites     D - actinolites

E - pyroxenites

44. When the margin connects segments between portions of a single spreading center _____ .

A - passive continental     B - continental convergent   C - continental collision

*D - transform fault continental     E - accreted terrane continental

45. A former margin of another type that has been modified by the addition of rafted-in, exotic fragments of crust such as island arcs _____ .

A - passive continental     B - continental convergent   C - continental collision

D - transform fault continental     *E - accreted terrane continental

46. The Atlantic Ocean margins of the Americas, Africa, and Europe are examples of _____ margins.

*A - passive continental     B - continental convergent   C - continental collision

D - transform fault continental     E - accreted terrane continental

47. The Andean coast of South America is an example of _____ margin.

A - passive continental     *B - continental convergent   C - continental collision

D - transform fault continental     E - accreted terrane continental

48. The line of collision between the Australian - Indian Plate and Asia is an example of a _____ margin.

A - passive continental     B - continental convergent   *C - continental collision

D - transform fault continental     E - accreted terrane continental

49. The western margin of North America from the Gulf of California to San Francisco is a _____ margin.

A - passive continental     B - continental convergent   C - continental collision

*D - transform fault continental     E - accreted terrane continental

50. The northwest margin of North America, from central California to Alaska is a _____ margin.

A - passive continental     B - continental convergent   C - continental collision

D - transform fault continental     *E - accreted terrane continental

51. A smaller fault bounded geologic entity characterized by a distinctive stratigraphic sequence and structural history that has become part of a larger continental mass is a _____ .

    A - captive     *B - terrane    C - alien      D - extraneous      E - collation

52. A body of fractured, deformed, thrust faulted sedimentary rocks that have been moved over a detachment surface is a _____ .

    A - detachee    *B - décollement      C - contraction block      D - voyageur

    E - vagrant

53. When the laterites of the Western Australian Shield formed, the region was probably in a _____ climate.

    A - wet tropical       B - dry subtropical       C - dry temperate

    *D - wet temperate       E - subpolar

## TRUE-FALSE

1. One of Alfred Wegener's principle lines of evidence for the existence of Pangaea was that ice sheets had covered parts of the southern continents.

    *T/F

2. When Alfred Wegener proposed his idea of continental drift all scientists accepted the idea with enthusiasm.

    T/*F

3. The apparent polar wandering path for each continent is different.

    *T/F

4. All that can be deduced from magnetic time lines the relative velocity of two plates.

    *T/F

5. Mélange are characterized by blue schists.

    *T/F

6. Transform faults form only at divergent plate margins.

    T/*F

7. More than 100 hot spots have been identified.

    *T/F

8. It is a well established fact that plate motions are caused entirely by convective motions in the mantle.

    T/*F

9. When the laterites of the Western Australian Shield formed, the region was probably in a temperate zone of high rainfall climate.

## COMPLETION

1. The idea that the Earth's poles have changed position through time is called **apparent polar wandering**.

2. The topography of the seafloor is controlled by the growth and movement of **plates**.

3. If the angle of plunge of a descending plate is very steep, the magmatic arc will have a **slight** curvature.

4. An assemblage of cratons and ancient orogens that has reached isostatic equilibrium is called a **continental shield** .

5. A smaller fault bounded geologic entity characterized by a distinctive stratigraphic sequence and structural history that has become part of a larger continental mass is a **terrane**.

6. The collision line between the two masses of deformed sediment caught up in a continental collision is commonly marked by the presence of **serpentinites**

## DISCUSSION

1. Discuss the three lines of paleomagnetic data that can be derived from rock studies.

2. Describe the test for seafloor spreading devised by Vine-Matthews-Morley.

3. Discuss the relationship between oceanic spreading centers, rate of heat loss, and general depth of the ocean.

4. Describe the formation, history and structure of a mélange.

5. Describe the origin of a back-arc basin.

6. Discuss the three mechanisms your text proposes for movement of the lithosphere.

7. Describe the tectonic-depositional history of the Red Sea.

8. Describe and discuss the origin and significance of terranes.

9. What are the two things suggested by the fact that the Western Australian Shield is weathered to a depth of hundreds of meters and what are the implications of this is relation to plate tectonics.

# CHAPTER 17

## RESOURCES OF MINERALS AND ENERGY

### MULTIPLE CHOICE

1. Probably the very earliest items traded by early man were _____ .

    A - salt        B - gold        *C - chert,flint, obsidian        D - copper        E - iron

2. Probably the earliest metals used by man were _____ .

    A - iron        B - gold        C - copper        D - only A and B        *E - only B and C

3. The first people to use oil instead of wood for fuel were the _____ .

    A - Chinese    B - Greeks        C - Egyptians  *D - Babylonians        E - Phoenicians

4. The first people to mine and use coal were the _____ .

    *A - Chinese    B - Greeks        C - Egyptians  D - Babylonians        E - Phoenicians

5. Today more than _____ kinds of minerals are mined and used.

    A - 150        B - 175        *C - 200        D - 225        E - 250

6. Any volume of rock containing an enrichment of one or more minerals is a _____ .

    A - ore body    B - mineral vein        C - mother lodd     *D - mineral deposit

    E - mine

7. The subfield of geology that deals with discovering new supplies of usable minerals is called _____ geology.

    A - pursuit    B - quest        C - analysis  D - research  *E - exploration

8. Mineral deposits are _____ .

    A - widespread but exhaustible        B - relatively rare but infinite

    C - widespread and infinite        *D - relatively rare and exhaustible

    E - none of the above

9. An aggregate of minerals from which one or more mineral(s) can be extracted profitably is a(n) _____ .

    *A - ore        B - lode        C - mine        D - quarry        E - excavation

10. The following are typical ore minerals.

A - sphalerite, galena, quartz          B - quartz, feldspar, calcite

C - chalcopyrite, dolomite, mica     D - dolomite, galena, mica

*E - chalcopyrite, galena, sphalerite

10. The following are typical gangue minerals.

A - sphalerite, galena, quartz          *B - quartz, feldspar, calcite

C - chalcopyrite, dolomite, mica     D - dolomite, galena, mica

E - chalcopyrite, galena, sphalerite

11. Concentration of mineral deposits by hot, aqueous solutions flowing through fractures and pore spaces in crustal rocks are _____ mineral deposits.

*A - hydrothermal     B - magmatic          C - sedimentary          D - placer

E - residual

12. Concentration by processes within a body of igneous rocks are _____ mineral deposits.

A - hydrothermal     *B - magmatic          C - sedimentary          D - placer

E - residual

13. Concentration by precipitation from lake water or seawater forms _____ mineral deposits.

A - hydrothermal     B - magmatic          *C - sedimentary          D - placer

E - residual

14. Concentration by flowing surface water in streams or along the shore to form _____ mineral deposits.

A - hydrothermal     B - magmatic          C - sedimentary          *D - placer

E - residual

15. Concentration by weathering processes to form _____ mineral deposits.

A - hydrothermal     B - magmatic          C - sedimentary          D - placer

*E - residual

16. It is probable that more mineral deposits have been formed by deposition from _____ than by any other mechanism.

A - hydrothermal solutions  B - magmatic solutions          C - sedimentary processes

D - placer processes          *E - residual processes

17. Hydrothermal solutions are associated with _____ .

A - volcanic calderas          B - granitic stocks and batholiths

C - high grade metamorphism          *D - both A and B          E - both A and C

18. Magmatic mineral deposits are formed principally by _____ .

    A - fractional crystallization          B - filter pressing          C - crystal setting

    D - both A and B          *E - both A and C

19. Pegmatites form by _____ .

    *A - fractional crystallization          B - filter pressing

    C - crystal setting          D - both A and B          E - both A and C

20. High density minerals such as chromite form by _____ .

    A - fractional crystallization          B - filter pressing

    *C - crystal setting          D - both A and B          E - both A and C

21. Typical evaporite minerals are _____ .

    A - halite, gypsum, quartz          B - quartz, feldspar, mica          C - pyroxene, galena, sylvite

    *D - halite, gypsum, borax          E - kaolinite, quartz, carnallite

22. Lake Superior-type iron deposits appear to have formed _____ .

    A - by hydrothermal solutions          B- magmatic processes

    C - evaporation from modern type oceans

    *D - evaporation from different than modern type oceans          E - none of the above

23. Lake Superior-type iron deposits may be turned into iron ores by ____ .

    A - hydrothermal leaching          B - leaching during weathering

    C - metamorphism          D - both A and B          *E - both B and C

24. During metamorphism of Lake Superior-type iron deposits _____ form.

    *A - magnetite and hematite          B - calcite and quartz

    C - magnetite and galena          D - galena and hematite          E - galena and quartz

25. Most stratabound mineral deposits are probably _____ in origin.

    A - magmatic *B - diagenetic          C - metamorphic          D - placer          E - residual

26. The most important minerals concentrated in placers are _____ .

    A - gold          B - platinum          C - diamond          *D - all of the above          E - only A and B

27. More than ____ of all the gold recovered throughout human history is from placers.

A - one-forth   B - one-third   *C - one-half   D - two-thirds        E - three-forths

28. Probably the richest gold placer deposits in the world is the_____ .

    A - Broken Hill in Australia        B - Kimberley in British Columbia

    C - Kupferschiefer in Germany        D - Hamersley Range in Australia

    *E - Witwatersrand Basin in South Africa

29. The most valuable residual mineral deposits are _____ .

    *A - bauxite   B - limonite   C - kaolinite    D - tripoli   E - caliche

30. A limited region of the crust within which mineral deposits occur in unusually large numbers is a _____ province.

    A - residual   B - depositional    C - metalliferous    *D - metallogenic

    E - porphyry

31. The geochemically abundant metals tend to be concentrated as _____ .

    A - oxides   B - hydrites   C - hydroxides    D - both A and B    *E - A and C

32. The geochemically abundant metals constitute more than _____ percent by weight of the Earth's crust.

    A - 0.01    B - 0.05    *C - 0.1    D - 0.5    E - 1.0

33. The geochemically scarce metals tend to be concentrated as _____ .

    A - oxides    *B - sulfides   C - hydroxides    D - both A and B    E - A and C

34. Most of the organics that become fossil fuels are trapped in _____ .

    A - sandstone   *B - shale    C - limestone   D - evaporites    E - dolostone

35. Coalification involves the loss of _____ materials.

    A - mineral   B - kerogen    C - lipids    *D - volatile   E - resins

36. Which of the following is most easily ignited?

    A - peat    *B - lignite    C - subbituminous coal    D - bituminous coal

    E - anthracite coal

37. Coal formed in _____ .

    A - slowly subsiding basins in continental interiors

    B - swampy margins of shallow inland seas at times of high sea level

    C - continental margins with wide continental shelves

    *D - all of the above    E - only A and C

38. Coal forms in _____ .

      *A - swamps    B - lakes      C - alluvial fans     D - stream channels  E - playas

39. The Pennsylvanian coals were formed mostly from _____ and the Cretaceous coals were formed mostly from _____ .

      A - bryophytes/angiosperms      B - ferns, gymnosperms/lycopods

      C - bryophytes/lycopods     *D - ferns, gymnosperms/ angiosperms

      E - angiosperms/ ferns, gymnosperms

40. Oil and gas form in _____ .

      A - sandstone      B - limestone      *C - shale

      D - A and B      E - A and C

41. Oil and gas are recovered from _____ .

      A - sandstone      B - limestone      C - shale

      *D - A and B      E - A and C

42. Approximately 60 percent of all oil and gas recovered is from _____age rocks.

      A - Pennsylvanian   B - Triassic      C - Cretaceous

      D - Mesozoic      *E - Cenozoic

43. Older rocks units do not produce as much oil and gas as recent ones because _____ .

      A - plants are very recent    B - conditions did not permit formation

      *C - oil escaped to the surface     D - no one knows how to find it

      E - the older rocks are too deep to drill into

44. Petroleum forms only if _____ .

      A - geothermal gradient is 1.8 to 5.5 C°/100 m     B - suitable source rock is present

      C - suitable traps are present     *D - all of the above     E - only A and B

45. Most oil and gas pools are found beneath _____ .

      A - faults    *B - anticlines    C - synclines

      D - monoclines     E - homoclines

46. Tar sands and oil shale are recovered by _____ .

      A - drilling   B - mining   C - heating   D - A and B   *E - B and C

47. It requires approximately _____ liters of oil to mine and process a ton of oil shale.

      A - 10        B - 20        C - 30      *D - 40      E - 50

48. The largest reserves of oil shale are in ____ .

   *A - Colorado          B - U.S.S.R.   C - Brazil       D - South Africa       E - China

49. Oil shale contains the compound ____ .

   A - oil          *B - kerogen   C - tar          D - gas          E - coal

50. If all of the hydroelectric potential on the plant was fully developed only ____ of the total needs for energy would be filled.

   A - one-forth          *B - one-third          C - one-half

   D - two-thirds          E - three-forths

51. For most countries, development of hydroelectric power has ____ potential.

   A - excellent   B - moderate          *C - limited    D - poor          E - nonexistent

52. To produce nuclear power, a controlled chain reaction must occur. The only naturally occurring material that will do so is ____ .

   A - $^{238}U$          *B - $^{235}U$          C - $^{233}U$          D - $^{232}Th$          E - $^{239}Pu$

53. Radioactive materials such as ____ may be produced in breeder reactors.

   A - $^{238}U$          B - $^{233}U$          C - $^{239}Pu$          D - both A and B          *E - both B and C

## TRUE-FALSE

1. Today more than 200 kinds of minerals are mined and used.

   *T/ F

2. Mineral and energy resources occur in virtually unlimited quantities and are easy to locate.

   T/*F

3. Once a mineral deposit has been declared an ore body it will always remain so.

   T/*F

4. Mineral deposits form by 5 processes, or combinations of processes.

   *T/ F

5. Which ore constituents are carried in hydrothermal solution depends on the kinds of rocks involved in the formation of the solution.

   *T/ F

6. Iron ores formed as a result of metamorphism are called taconites.

   *T/ F

7. A limited region of the crust within which mineral deposits occur in unusually large numbers is a metallogenic province.

*T/ F

8.Energy utilization is uniform worldwide.

T/*F

9. Experts agree closely on the amount of oil and gas remaining to be recovered.

T/*F

10. Biomass energy has the potential for much greater development without hurting the environment.

*T/ F

11. Nuclear reactors use the heat produced by nuclear fusion to operate.

T/*F

12. Geothermal energy offers the greatest potential for solving the energy needs on mankind.

T/*F

## COMPLETION

1. It is common practice to restrict use of the term sedimentary to those mineral deposits formed through **precipitation** of substances carried in solution.

2. Most stratabound mineral deposits are probably **diagenetic** in origin.

3. The five geochemically abundant metals in the Earth's crust are: **aluminum, iron, manganese, magnesium, titanium**.

## DISCUSSION

1. Discuss the three aspects of minerals as your text dealt with them.

2. What causes a mineral deposit to become an ore body and an ore body to become just a mineral deposit.

3. Describe the history of hydrothermal solutions that form volcanogenic massive sulfide deposits.

4. Describe the geology of the Witwatersrand Basin in South Africa.

5. Discuss the difference is source material for coal vs. oil and natural gas.

6. Describe the sequence of events in the coalification process.

7. Discuss the potential for future development of biomass energy.

8. Discuss the problems involved with generating power using nuclear energy.

9. Discuss the three ore formation sites discovered since the 1960's.

# CHAPTER 18

## BEYOND PLANET EARTH

### MULTIPLE CHOICE

1. There are _____ planets in this solar system with rings.

    A - 1        B - 2        C - 3        *D - 4        E - 5

2. The Greek name *planetai* for the planets means _____ .

    *A - wanderers        B - travelers    C - strangers    D - quests    E - rovers

3. The terrestrial plants have densities greater than_____ $g/cm^3$.

    A - 2.0        B - 2.5        *C - 3.0        D - 3.5        E - 4.0

4. The jovian planets have densities less than_____ $g/cm^3$.

    A - 0.5        B - 1.0        C - 1.5        *D - 2.0        E - 2.5

5. All of the planets have in common _____ .

    A - nearly the same plane of revolution        B - nearly the same size

    C - same direction of revolution        D - both A and B        *E - both A and C

6. The name of the Earth's orbit around the sun is the _____ .

    A - elliptic        B - plane        C - plane of revolution        D - plane of rotation

    *E - ecliptic

7. All of the planets appear to have a _____ .

    A - biosphere    B - hydrosphere        C - moon        *D - solid body        E - icy layer

8. The _____ Rule states that each planet in the solar system is approximately twice as far from the sun as the next one closer to the sun.

    A - Skinner-Porter    B - Hutton-Lyell        C - Jones-Laughlan

    D - Schmidt-Longacre        *E - Titius-Bode

9. The space between Mars and Jupiter is occupied by _____ .

    A - bolides    *B - asteroids    C - meteorites        D - capsularia        E - ejecta

10. Approximately _____ impact craters have been discovered on Earth.

    A - 50        B - 100        C - 150        *D - 200        E - 250

11. Relatively few impact craters have been discovered on Earth because _____ .

    A - most rocks are too resistant to crater    B - the gravitational field is too weak

    *C - weathering masks them    D - bolide are very rare in this region

    E - Jupiter is too large

12. Volcanism is predominantly _____ on the terrestrial planets and moons.

    A - rhyolitic        B - andesitic        *C - basaltic

13. The chemical-mineralogical composition of the terrestrial planets and moons appears to be _____ .

    *A - very similar    B - similar    C - slightly similar

    D - dissimilar    E - very dissimilar

14. Most of what we know about the planet Mercury is from _____ .

    A - telescopic observations    B - unmanned lander missions

    C - manned lander missions    *D - unmanned fly-by missions

    E - manned fly-by missions

15. Images of Mercury's surface suggests that _____ .

    *A - it lacks an atmosphere    B - the surface is smooth

    C - plate tectonics is active    D - all of the above    E - only B and C

16. The metallic core of Mercury is _____ the size of the planet.

    *A - almost equal to    B - three-quarters    C - one-half

    D - one-third    E - one-forth

17. Comparing the Earth-moon system to other planet-moon systems, Earth's moon is _____ .

    A - very small    B - small    C - moderate  D - large    *E - very large

18. The most informative measurements made by astronauts of the moon's structure were _____ .

    A - type of rock on the maria    B - size of impact craters    *C - seismic waves

    D - type of rock in the uplands    E - distribution pattern of craters

19. The moon has fewer than _____ moonquakes per year.

    A - 500        *B - 400        C - 300        D - 200        E - 100

20. Moonquakes are probably caused by _____ .

    *A - Earth's gravitational pull    B - plate tectonics

C - volcanism          D - only A and B          E - only B and C

21. We know the moon's lithosphere is at least 1,000 km thick because of _____ .

   *A - the density of the moon          B - the types of rocks that occur

   C - the size of the craters          D - earthquakes 1,000 km deep

   E - the type of volcanic activity

22. The oldest rocks from the moon are _____ , nearly 4.5 billion years old.

   *A - anorthosite(Ca-rich plagioclase)          B - granite( K-rich feldspar)

   C - basalt(K and P rich)          D - basalt(Fe and Ti rich)          E - rhyolite(K-rich)

23. The second oldest rocks from the moon are _____ , nearly 4.0 billion years old.

   A - anorthosite(Ca-rich plagioclase)          B - granite( K-rich feldspar)

   *C - basalt(K and P rich)          D - basalt(Fe and Ti rich)          E - rhyolite(K-rich)

24. The oldest rocks from the moon are _____ , nearly 3.2 to 3.8 billion years old.

   A - anorthosite(Ca-rich plagioclase)          B - granite( K-rich feldspar)

   C - basalt(K and P rich)          *D - basalt(Fe and Ti rich)          E - rhyolite(K-rich)

25. The moon's surface is covered by _____ .

   A - gray pulverized rock fragments          B - glassy dust

   C - micrometeorites          *D - both A and B          E - both A and C

26. Mars is similar to the Earth in that it _____ .

   A - rotates each 24.6 hours          B - has an atmosphere

   C - has polar ice caps          *D - all of the above          E - none of the above

27. Mars appears to lack _____ .

   A - a dipolar magnetic field          B - marsquakes

   C - volcanic activity          *D - only A and B          E - only B and C

28. The southern hemisphere of Mars is _____, while the northern hemisphere has striking _____ .

   A - river valleys/impact craters          B - volcanos/ glacial features

   *C - impact craters/volcanos          D - river valleys/volcanos

   E - glacial features/impact craters

29. Mars has evidence of erosion by _____ .

A - running water     B - wind     C - glaciers     *D - only A and B     E - B and C

30. The *Viking 1 and 2* landers found the surface of Mars covered with _____ .

    *A - clays and a sulfate mineral     B - clays and a sulfide mineral

    C - carbonate and sulfate minerals     D - clays and carbonate minerals

    E - sulfate and sulfide minerals

31. The southern hemisphere of Mars appears to be similar to _____ .

    A - Asia     *B - the Moon     C - Venus     D - Mercury     E - Jupiter

32. Early in martian history, it may have been covered by a layer of water _____ m thick.

    A - 25 to 50     B - 25 to 75     C - 50 to 75     *D - 50 to 100     E - 75 to 100

33. The martian water is probably tied up as _____ .

    A - cave fillings     B - ice caps     *C - permafrost     D - lost to space

    E - the above are all ridiculous

34. Venus is the planet most like the Earth in _____ .

    A - size     B - atmosphere     C - mass     D - both A and B     *E - A and C

35. Venus has an atmosphere dominated by _____ .

    *A - carbon dioxide     B - water vapor     C - nitrogen     D - methane     E - ammonia

36. The surface temperature of Venus is approximately _____ °C.

    A - 350     B - 400     C - 450     *D - 500     E - 550

37. The surface of Venus is unimodal suggesting that the planet is _____ .

    A - active plate tectonics     *B - nonactive plate tectonics

    C - volcanically active     D - volcanically inactive     E - it does not mean anything

38. The surface of Venus is dotted with _____ .

    A - basaltic volcanos     B - rhyolitic volcanos     C - volcanic plains

    *D - all of the above     E - none of the above

39. Approximately _____ percent of Venus is covered by volcanic plains.

    A - 50     B - 60     C - 70     *D - 80     E - 90

40. The atmospheres of Earth and Venus differ largely because _____ .

    A - they were originally different     B - distance from the sun

    C - Venus has more moons     D - their mass is different     *E - life evolved on Earth

41. Jupiter is peculiar in that it _____ .

   A - is retrograde          B - is twice as massive as all the other planets

   C - gives off twice as much energy as it receives     D - all of above     *E - B and C

42. The most volcanically active body in the solar system is _____ .

   A - Venus      B - Mars      C - Earth      D - Europa          *E - Io

43. Volcanism on Io is _____ .

   A - basaltic    B - rhyolitic   C - sulfureous        D - all of the above    *E - A and C

44. Io appears to derive most of its energy from _____ .

   *A - gravitational attractional forces          B - radioactive decay

   C - gravitational collapse forces     D - both A and B      E - both B and C

45. As moons get further from Jupiter, the density _____ and the ice layer thickness _____ .

   A - decreases/decreases          B - increases/increases

   C - increases/decreases     *D - decreases/increases      E - no change occurs

46. The only moon in the solar system large enough to have an atmosphere is _____ .

   A - Io          *B - Titan      C - Ganymede        D - Callisto    E - Europa

47. Saturn's moon Titan may have oceans of _____ and continents of _____ .

   A - water/rock       B - water/ice   C - liquid hydrocarbons/rock

   D - water/frozen hydrocarbons     *E - liquid hydrocarbons/ ice

## TRUE-FALSE

1. The origin of the rings around Saturn is easily explained.

   T/*F

2. Asteroids and meteoroids are probably objects captured from outer space.

   T/*F

3. The origin of Mercury's magnetic field is very difficult to explain.

   *T/F

4. The density of Earth and Mars is very similar.

   *T/F

5. Olympus Mons, on Mars, has a greater volume than all of the Hawaiian Islands combined.

   *T/F

6. The more we learn about Venus, the more apparent it becomes that Venus and the Earth are more similar than the Earth and any other planet.

   *T/F

7. The Earth has locked up most of its carbon in the form of living biomass.

   T/*F

8.The composition of Saturn and Jupiter is similar.

   *T/F

## COMPLETION

1. All of the planets except **Uranus, Neptune, and Pluto** have been known since antiquity.

2. The shape of the orbit of each planet is **elliptical**.

3. Impact craters form when a **bolide** impacts a planetary body.

4. The moon appears to have grown by accretion of **bolides**.

## DISCUSSION

1. Discuss the sequence of events when a high-speed bolide impacts and penetrates the surface of a planet.

2. What are the 6 pieces of information about the structure of the moon that have been determined from the study of moonquakes?

3. Describe the accretionary-tectonic history of the moon.

4. Describe the "Geological History" of Mars.

5. Compare the Terrestrial Planets.

6. Discuss Saturn's moon Titan.

# CHAPTER 19

## OUR CHANGING PLANET

### MULTIPLE CHOICE

1. Ozone in the atmosphere serves the function of _____ .

    A - breaking down carbon monoxide     B - stabilizing temperatures

    C - reflecting solar radiation     *D - absorbing ultraviolet radiation

    E - breaking down chlorofluorocarbons

2. Ozone has the following formula _____ .

    A - O     B - $O_2$     *C - $O_3$     D - $O_4$     E - $O_5$

3. Chlorofluorocarbons pose the problem of _____ .

    A - breaking down carbon monoxide     B - forming carbon monoxide

    C - absorbing ultraviolet radiation     D - reflecting solar radiation

    *E - breaking down ozone

4. The really damaging part of chlorofluorocarbons is the _____ that forms during breakdown.

    *A - Chlorine     B - Fluorine     C - Carbon

    D - Nitrouschlorine    E - Nitrousfluoride

5. Ozone breakdown becomes most serious at temperatures below − _____ °C.

    A - 40     B - 50     C - 60     D - 70     *E - 80

6. A single chlorine atom can destroy as many as _____ ozone molecules before it is chemically bound-up.

    A - 100     B - 1,000     C - 10,000     *D - 100,000    E - 1,000,000

7. Naturally occurring events that cause global changes are such as _____ .

    A - very large volcanic eruption     B - very large meteorite impact

    C - large earthquake     *D - both A and B     E - both A and C

8. During a typical lifetime a crustal plate will move approximately _____ meters.

    A - 2     B - 5     C - 7     *D - 10     E - 12

9. The theory of plate tectonics allows us to deal with questions such as _____ .

    A - How are mountains formed?    B - How does the weather change?

    C - Why are the Great Lakes present?    *D - both A and C    E - both A and B

10. The theory of evolution allows us to deal with questions such as _____ .

    *A - Why are there so many different organisms?    B - Why are fossils in rocks

    C - Why do ice ages occur?    D - both A and B    E - both A and C

11. The unified theory of climate allows us to deal with questions such as _____ .

    A - Why do ice ages occur?    B - Why are reefs in rocks in the Arctic?

    C - Has the climate of the Earth changed during time?

    D - all of the above    *E - none of the above

12. Carbon occurs in the atmosphere as _____ .

    *A - carbon dioxide    B - organic compounds    C - free carbon

    D - dissolved carbon dioxide    E - calcium carbonate

13. Carbon occurs in the biosphere as _____ .

    A - carbon dioxide    *B - organic compounds    C - free carbon

    D - dissolved carbon dioxide    E - calcium carbonate

14. Carbon occurs in the hydrosphere as _____ .

    A - carbon dioxide    B - organic compounds    C - free carbon

    *D - dissolved carbon dioxide    E - calcium carbonate

15. Carbon occurs in the crust as _____ .

    A - carbon dioxide    B - organic compounds    C - free carbon

    D - dissolved carbon dioxide    *E - calcium carbonate

16. The primary realm for cycling carbon compounds is the _____ .

    A - atmosphere    B - hydrosphere    *C - biosphere

    D - crust    E - lithosphere

17. The entire content of carbon dioxide in the atmosphere is cycled through the biosphere in only _____ years.

    A - 1.5    B - 2.5    C - 3.5    *D - 4.5    E - 5.5

18. The carbon dioxide content of the atmosphere will be increased by _____ .

A - planting vegetation    *B - burning fossil fuels    C - cooling the oceans

D - all of the above    E - none of the above

19. The carbon dioxide content of the atmosphere will be decreased by _____ .

A - planting vegetation    B - burning fossil fuels    C - cooling the oceans

D - all of the above    *E - only A and C

20. The atmosphere readily allows _____ radiation to pass through.

*A - short wave length    B - long wave length

21. The phenomenon of the atmosphere capturing heat is the _____ effect.

A - thermal    B - Kelvin    C - atmospheric dynamics

D - pop can    *E - greenhouse

22. The most important gas in the atmosphere contributing to the greenhouse effect is _____ .

*A - water vapor    B - carbon dioxide    C - methane

D - nitrous oxide    E - ozone

23. The second most important gas in the atmosphere contributing to the greenhouse effect is _____ .

A - water vapor    *B - carbon dioxide    C - methane

D - nitrous oxide    E - ozone

24. The concentration of carbon dioxide in the atmosphere is increasing at a _____ rate.

A - arithmetic    B - linear    *C - exponential

D - geometric    E - none of the above

25. Methane absorbs ultraviolet radiation approximately _____ times as efficiently as carbon dioxide.

A - 5    B - 10    C - 15    D - 20    *E - 25

26. The principal sources of methane in the atmosphere are _____ .

A - biological activity related to rice cultivation

B - leaks in domestic and industrial gaslines

C - byproduct of the digestive processes of domestic livestock

*D - all of the above    E - only B and C

27. The concentration of methane in the atmosphere is increasing approximately 1% per year because: _____ .

A - more land is being put into rice cultivation

B - there are increasing numbers of livestock

C - there are more gasline leaks

*D - all of the above          E - only A and C

28. Chlorofluorocarbons are approximately ____ times as efficient as carbon dioxide at absorbing ultraviolet radiation.

    A - 1,750      B - 2,000      C - 10,000      D - 15,000      *E - 20,000

29. Chlorofluorocarbons concentrations in the atmosphere are increasing at approximately ____ per cent per year.

    A - 1      B - 2      C - 3      D - 4      *E - 5

30. Ozone in the troposphere is produced largely by ____ .

    *A - combustion of fossil fuels      B - decay of agricultural residues

    C - soil microbial activity      D - all of the above    E - only B and C

31. Nitrous oxide in the troposphere is produced largely by ____ .

    A - combustion of fossil fuels      B - decay of agricultural residues

    C - soil microbial activity    *D - all of the above      E - only B and C

32. Global temperature curves used to depict global temperature rises are ____ .

    A - mainly for the northern hemisphere      B - for the continents

    C - based upon data for several centuries    D - all of the above    *E - only A and B

33 . Global climate models attempt to link processes in the ____ .

    A - atmosphere      B - hydrosphere      C - biosphere

    *D - all of the above    E - only A and B

34. Global climate models do not adequately portray ____ .

    A - dynamics of ocean circulation    B - dynamics of cloud formation

    C - the effects of lithospheric dynamics    D - all of the above    *E - only A and B

35. Global climate models have been ____ successful in simulating the general character of present-day climates.

    *A - very      B - moderately      C - slightly    D - poorly    E - totally un

36. One of the primary controls on operating global climate models is ____ .

    A - amount of data for input      B - amount of computer memory

    *C - cost of operating the computer    D - lack of trained operators

E - lack of interest by funding agencies

37. Five global climate models have been used most extensively for global warming predictions, the consensus of the models is that the average temperature increase will be _____ °C.

A - 0.25 to 0.75    *B - 0.5 to 1.5   C - 1.0 to 3.0   D - 2.0 to 4.0   E - 3.0 to 5.0

38. Five global climate models have been used most extensively for global warming predictions, the consensus of the models is that the average temperature increase will not be uniform and the greatest increases will be in _____ regions.

A - tropical    B - subtropical        C - temperate        D - desert       *E - polar

39. The difference in mean temperatures today and during the height of the ice ages is approximately _____ °C.

A - 2          B - 3          C - 4          *D - 5          E - 6

40. As global warming progresses precipitation in the tropics will probably _____ and in the centers of large continents will probably.

A - increase/increase              *B - increase/decrease

C - decrease/decrease        D - decrease/increase        E - no change anywhere

41. As global warming progresses, glaciers in the lower latitudes will probably _____ while those in high latitudes will probably.

A - increase/increase              B - increase/decrease

C - decrease/decrease        *D - decrease/increase        E - no change anywhere

42. As global warming progresses, temperatures in the northern high latitudes will probably _____ while those in southern high latitudes will probably.

A - increase/increase              B - increase/decrease

C - decrease/decrease        D - decrease/increase        *E - increase/little change

43. Gas hydrates are estimated to hold as much as 10,000 billion metric tons of carbon, an amount equal to _____ times all of the carbon held in the worlds coal, gas, and oil reserves.

*A - 2          B - 3          C - 4          D - 5          E - 6

44. Gas hydrates accumulate principally _____ .

A - in oceans below 500 m          B - beneath frozen ground

C - in association with clay deposits        *D - both A and B        E - both A and C

45. The statement, "Anything that did happen, can happen." is known as _____ Law.

A - Murphey's        B - Guy's        *C - Ayer's        D - Lyell's        E - Sorensen's

46. In the past global warming has caused _____ .

A - organisms to migrate poleward          B - sea level to rise

C - soil formation to go on at higher rates    *D - all of the above    E - none of the above

47. Global warming appears to be only temporary because _____ .

        A - new atmosphere is constantly forming        B - fossil fuel will be exhausted

        C - the geometry of Earth's orbit changes constantly

        D - both A and B        *E - both B and C

48. Evidence that mid-Cretaceous climates were considerably warmer than today are _____ than today.

        A - reef 5° to 15° further poleward    B - sea level 100 m to 200 m higher

        C - isotopic measurements        *D - all of the above    E - only A and B

49. Mid-Cretaceous climates were considerably warmer than today probably because there was much more _____ .

        A - carbon dioxide in the atmosphere        B - volcanic activity

        C - solar radiation    *D - both A and B    E - both B and C

50. The suggested source for the mid-Cretaceous volcanism is _____ .

        A - the Andean Mountain chain    *B - Ontong-Java Plateau

        C - Carlsberg Ridge        D - Emperor Seamounts    E - Tuamotu Archipelago

## TRUE-FALSE

1. When man interferes with a natural system the results are easily predicted.

    T/*F

2. The average person is well aware of the motion of a crustal plate.

    T/*F

3. Not all of the carbon compounds that occur in the biosphere are oxidized into carbon dioxide that goes into the atmosphere.

    *T/F

4. The rate oceans absorb carbon dioxide is relatively fixed.

    *T/F

5. If the Earth lacked an atmosphere the diurnal temperature range would be much greater.

    *T/F

6. Chlorofluorocarbons are approximately 20,000 times as efficient as carbon dioxide at absorbing ultraviolet radiation.

    *T/F

7. The most efficient agent for removing ozone from the troposphere is the photosynthetic processes of the tropical rain forests.

   *T/F

8. Evidence for a global temperature increase is absolutely conclusive.

   T/*F

9. Scientists are in strong agreement on what final effect global warming will have on the climate.

   T/*F

10. Superplumes probably cool the core of the Earth.

   *T/F

# DISCUSSION

1. Discuss 3 types of naturally occurring events that will indicate to the average person that the Earth is a dynamically changing planet.

2. Discuss the role of the several fields of geology in the development of a unified theory of climate.

3. Describe the carbon cycle.

4. Describe how the greenhouse effect works.

5. What are the reasons for the great rise in the atmospheric concentration of carbon dioxide during the last 200 years?

6. What are the three principal sources of methane in the atmosphere.

7. Discuss the typical boundary conditions for a global climate model.

8. Discuss the effect of breakdown if gas hydrates on the problems of global warming-as both cause and effect.

9. Discuss the mid-Cretaceous temperature rise - reasons, results, causes, etc.

# Part III

# SUPPLEMENTARY SLIDE DESCRIPTIONS

# SUPPLEMENTARY SLIDE DESCRIPTIONS

## INTRODUCTION
(Slides 1–6)

1. **FIGURE I.6.** The African Rift Valley, which extends from the Red Sea in the north to Malawi in the south, is a gigantic rent in the Earth's surface marking the place where deep-seated convection currents are splitting Africa into two pieces. This LANDSAT image is of a portion of the Rift Valley in central Kenya that is 70 km wide and 140 km long. To the east (righthand side) is a high plateau bounded by a series of north–south fractures that mark the edge of the Rift Valley. Several volcanoes are visible in the Valley.

**2 and 3.** **FIGURE I.7.** The internal structure of sand dunes, ancient and modern. **A.** A distinctive pattern of wind-deposited sand grains can be seen in a hole dug in this dune near Yuma, Arizona. **B.** The same distinctive pattern in rocks in Zion National Park, Utah, lets us infer that these rocks were once sand dunes too.

4. **FIGURE I.8.** Siccar Point, Berwickshire, Scotland. Layers of sedimentary rock, originally horizontal, were bent and tilted into vertical layers during uplift. Erosion developed a new land surface that, on submergence, became the surface on which younger sediments were laid. The eroded vertical layers are about 450 million years old. The layers above the old surface of erosion, which are named the Old Red Sandstone, are 370 million years old. At this locality, in 1788, James Hutton first demonstrated that the cycle of sedimentation, uplift, and erosion is repeated again and again.

5. **FIGURE I.9.** Meteor Crater, near Flagstaff, Arizona. The crater was created by the impact of a meteorite about 20,000 years ago. It is 1.2 km in diameter and 200 m deep. Note the raised rim of the crater wall and the blanket of broken rock debris thrown out of the crater. Many impacts larger than the meteor crater event are believed to have occurred during the Earth's long history.

6. **FIGURE I.10.** This thin, dark layer of rock (marked by the coin) is rich in the rare chemical element iridium and looks out of place in the thick sequence of pale-colored limestones above and below. The iridium-rich layer, here seen in the Contessa Valley, Italy, has been identified at many places around the world and is believed to have formed as a result of a world-encircling dust cloud formed by a great meteorite impact about 65 million years ago.

# CHAPTER 1
## (Slides 7–10)

**7 and 8. FIGURE 1.5.** Two moons of Jupiter that are rich in volatile elements. **A.** Io is red colored because it is rich in sulfur. The image shows a volcanic eruption on Io. The volcanic plume is mostly gas, but small solid particles are also distributed by the gas. The plume rises to a height of 100 km above the surface of Io and is believed to be largely sulfur dioxide ($SO_2$). Several sites of active volcanism have been discovered on Io. **B.** The surface of Europa, smallest of the four large moons of Jupiter, is mantled by ice to a depth of 100 km. The fractures indicate that some internal process, probably convection, must be renewing the surface. The dark material in the fractures apparently rises up from below. The cause of the fracturing is not known. The image was taken by *Voyager 2* in July 1979.

**9. FIGURE 1.14.** Long, deep fractures that split Iceland mark the center of a midocean ridge. Iceland is on the Mid-Atlantic Ridge and is one of the few places in the world where the midocean ridge can be seen above sealevel.

**10. FIGURE 1.20.** This chain of volcanoes in Ecuador sits above the subduction zone where tha Nazca Plate sinks below the western edge of the South American Plate. Seven snowcapped volcanoes are visible in this aerial photograph.

# CHAPTER 2
## (Slides 11–16)

**11. FIGURE 2.7.** Atoms can be seen through special kinds of microscopes. Sulfur atoms (large) and lead atoms (small) at the surface of a galena (PbS) crystal are revealed with a scanning-tunneling microscope.

**12. FIGURE 2.10.** Two quartz crystals with the same crystal forms. Although the size of the individual faces differ markedly between the two crystals, it is clear that each numbered face on one crystal is parallel to an equivalent face on the other crystal. It is a fundamental property of crystals that, as a result of the internal crystal structure, the angles between adjacent faces are identical for all crystals of the same mineral.

**13. FIGURE 2.14A.** Relation between crystal structure and cleavage. Halite, NaCl, has well-defined cleavage planes; it breaks into fragments bounded by perpendicular faces. **B.** The crystal structure in the same orientation as the cleavage fragments shows that the directions of breakage are planes in the crystal between equal numbers of sodium and chlorine atoms.

**14. FIGURE 2.16.** Two common minerals with distinct cleavages. Fluorite (left) breaks along four planes into octahedral cleavage fragments. Potassium feldspar (right) cleaves along the two planes that are perpendicular. The fluorite is from England, the feldspar from Maine.

**15. FIGURE 2.18.** Color contrast between hematite and a hematite streak. Massive hematite is opaque, has a metallic luster, and appears black. On a porcelain plate, hematite gives a red streak.

**16. FIGURE 2.27.** Calcite ($CaCO_3$) on the left, and dolomite $CaMg(CO_3)_2$ on the right have similar crystal structures and, as a result, similar cleavages. Both cleave along three planes that are not perpendicular, yielding rhombohedral-shaped fragments. The calcite is from Iceland, the dolomite from Traversella, Italy.

# CHAPTER 3
## (Slides 17–29)

**17. FIGURE 3.2.** An advancing tongue of basaltic lava setting fire to a house in Kalapana,

Hawaii, during an eruption of Kilauea volcano in June 1989. Flames at the edge of the flow are due to burning lawn grass.

18. **FIGURE 3.3.** The way lava flows is controlled by viscosity. Two different flows are visible. They have the same basaltic composition. The lower flow, on which the geologist is standing, is a pahoehoe flow formed from a low-viscosity lava like that shown in Figure 3.2 (Supplementary Slide 18). The upper flow (the one being sampled by the geologist), which is very viscous and slow moving, is an aa flow erupted from Kilauea volcano in 1989. The pahoehoe flow was erupted in 1959.

19. **FIGURE 3.4.** Spectacular fountaining starts an eruption of a basaltic volcano in Iceland. Use of a telephoto lens foreshortens the field of view. The geologists who are measuring the height of the fountain (200 m) are many hundreds of meters away from the erupting lava.

20. **FIGURE 3.5.** This stream of low-viscosity, basaltic lava moving smoothly away from an eruptive vent demonstrates how fluid and free flowing lava can be. The temperature of the lava is about 1100°C. The eruption occurred in Hawaii in 1983.

21. **FIGURE 3.6.** Vesicular basalt. Note the grains of yellowish-green olivine. The specimen is 4.5 cm across.

22. **FIGURE 3.10.** Mauna Kea, a 4200 m-high shield volcano on Hawaii, as seen from Mauna Loa. Note the gentle slopes formed by highly fluid basaltic lava. The view is almost directly north. A pahoehoe flow is in the foreground on the northeast flank of Mauna Loa.

23 and 24. **FIGURE 3.11.** Tephra cones. **A.** Two small tephra cones forming as a result of an eruption in Kivu, Zaire. Arcs of lights are caused by the eruption of red-hot lapilli and bombs. **B.** Tephra cone in Arizona built from lapilli-sized tephra. Note the small basaltic lava flow coming from the base of the cone.

25. **FIGURE 3.15.** Crater Lake, Oregon, occupies a caldera 8 km in diameter that crowns the summit of a once lofty stratovolcano, posthumously called Mount Mazama. Wizard Island, a small tephra cone, formed after the collapse that created the caldera.

26. **FIGURE 3.16.** The Pinnacles, Crater Lake National Park. Striking erosional forms developed in the thick tephra blanket left by the eruption of Mount Mazama 6600 years ago.

27. **FIGURE 3.18.** A lava dome in the crater of Mount St. Helens, Washington, in May 1982. The plume rising above the dome is steam.

28. **FIGURE 3.19.** Aerial view of a fissure eruption, Mauna Loa, Hawaii, in 1984. Basaltic lava is erupting from a series of parallel fissures. Note the tephra cone (upper left) formed during an earlier eruption.

29. **FIGURE 3.25.** Three coarse-grained igneous rocks. Compare their mineral assemblages by using the chart in Figure 3.24. Note the change in color from granite (left), which is light colored because it is rich in feldspar and quartz, through diorite (center), to gabbro (right), which is quartz free and rich in pyroxene and olivine and therefore darker in color. Each specimen is 7 cm across.

# CHAPTER 4
## (Slides 30–38)

30. **FIGURE 4.6.** Ancient cross-stratified sand dunes that have been converted to sedimentary rock that crops out near Kanab, Utah. Cross-strata are inclined to the right, in the direction toward which the ancient prevailing winds were blowing.

31. **FIGURE 4.7.** A mudflow deposit that originated at Mount St. Helens volcano is vertically sorted. Conspicuous cobbles and pebbles in the lower part grade upward into finer pebbles and sand near the top. This graded bed resulted from a prehistoric erup-

tion that sent a flood of muddy debris down a nearby valley.

32. **FIGURE 4.13B.** Irregular layers of black bedded chert occur in a thick section of fossiliferous light-gray limestone in the central Brooks Range of Alaska. The chert was used by prehistoric inhabitants of these mountains to make knives, points, and other tools.

33. **FIGURE 4.14.** Iron-rich sediments of the Brockman Iron Formation in the Hamersley Range of Western Australia. The white layers are largely chert, while the darker bluish and reddish layers consist mainly of iron-rich silicate, oxide, and carbonate minerals.

34. **FIGURE 4.15.** Evaporite salts encrust the surface of a desert playa on the floor of Death Valley, California. A shallow lake forms during rainy periods. As the water then evaporates and the playa lake dries up, salts crystallize out of the brine, and polygonal fractures form as the drying sediment contracts.

35 and 36. **FIGURE 4.20.** Modern and ancient mudcracks. **A.** Mudcracks formed the surface of this dry lake floor. **B.** Ancient mudcracks preserved on the surface of a mudstone bed exposed at Ausable Chasm, New York.

37. **FIGURE 4.25.** Carbonate sediments consisting of fine skeletal debris and inorganic precipitates accumulate in the warm, shallow marine waters of a broad, flat carbonate shelf surrounding the numerous islands of Bermuda.

38. **FIGURE 4.28.** Deep-sea turbidite beds that have been tilted, uplifted, and exposed in a wave-eroded bench along the coast of the Olympic Peninsula, Washington.

# CHAPTER 5
## (Slides 39–45)

39 and 40. **FIGURE 5.2.** Comparison of textures developed in rocks of the same composition under uniform and differential stress. **A.** Granite, consisting of quartz, feldspar, and biotite that crystallized in a uniform stress field. Note that biotite grains are randomly oriented. **B.** High-grade metamorphic rock, also consisting of quartz, feldspar, and biotite, that crystallized in a differential stress field. Biotite grains are parallel, giving the rock a distinct foliation.

41. **FIGURE 5.3.** Deformation of a conglomerate during metamorphism. Sandstone pebbles, originally round, have been flattened. The direction of maximum stress is indicated by arrows. The pocket knife is 6 cm long.

42. **FIGURE 5.12.** Cataclastic metamorphism of a granitic dike intruded into a biotite-rich gneiss. Brittle granite, rich in feldspar (pink) and quartz (light gray), was fragmented during deformation. The gneiss did not fragment because it was ductile and tended to flow plastically. The field of view is about 50 cm across.

43 and 44. **FIGURE 5.13.** Development of foliation in a granite by cataclastic metamorphism. From Groothoek, South Africa. **A.** Undeformed granite consisting of quartz, feldspar, and biotite. The dark patch in the center of the field of view is a xenolith of amphibolite. Foliation is not present in the granite. **B.** The original granitic texture has been completely changed, and the granite has been transformed to a gneiss with a distinct foliation. Amphibolite xenoliths have been flattened and elongated.

45. **FIGURE 5.21A.** Hydrothermal solution forming an ore deposit on the seafloor. A so-called "black smoker" photographed at a depth of 2500 m below sea level on the East Pacific Rise at 21°N latitude. The "smoker" has a temperature of 320°C. The rising hydrothermal solution is actually clear; the black color is due to fine particles of iron sulfide and other minerals precipitated from solution as the plume is cooled through contact with cold seawater. The chimneylike structure is composed of pyrite, chalcopyrite and other ore minerals deposited by the hydrothermal solution.

# CHAPTER 6
## (Slides 46–47)

**46. FIGURE 6.1.** The Grand Canyon of the Colorado River. Flat-lying strata, nearly 2000 m thick and accumulated over 300 million years, were laid down on a basement of tilted and tectonically deformed igneous and metamorphic rocks.

**47. FIGURE 6.6.** Spectacular view of the Navajo Sandstone, Zion National Park, Utah. The Navajo Sandstone is an example of a rock-stratigraphic unit.


# CHAPTER 7
## (Slides 48–59)

**48. FIGURE 7.1.** A roadcut on the island of Hawaii exposes a weathering profile through a thick lava flow. Loose, earthy regolith grades downward into discolored, weathered rock that retains its organized appearance. Fresh, virtually unweathered rock is seen at the base of the exposure near the figure.

**49. FIGURE 7.2.** Sheetlike jointing in massive granite forms stair-step surface on a mountainside in Yosemite National Park, California.

**50. FIGURE 7.3.** Three sets of joints, one horizontal and two vertical, intersect at nearly right angles to form a spectacular rocky vantage point, called the Pulpit, overlooking Lysefjord in southwestern Norway. A widening crack along a vertical joint suggests that the Pulpit eventually will collapse and plunge into the icy waters of the fjord far below.

**51. FIGURE 7.4.** Columnar jointing in igneous rock near San Miguel Regla, Mexico, offers a challenge to a rock climber jambing his way up a crack between two adjacent columns.

**52. FIGURE 7.7.** A large forest fire, moving rapidly through a pine forest in Yellowstone National Park, Wyoming, has caused extensive spalling of a surface boulder. Fresh, light-colored spalls, which flaked off as heat caused differential expansion of the rock surface, litter the fire-blackened ground.

**53. FIGURE 7.11.** Exfoliating granite boulders near Iferouane in the Air Mountains of Niger. Thin, sheetlike spalls flake off the rock as it weathers, gradually causing the boulders to increase in sphericity.

**54. FIGURE 7.12.** Spheroidal weathering of granite bedrock in the northern Sierra Nevada of California produces boulders of solid granite that are surrounded by a core of disintegrated rock. Although the boulders resemble rounded stream gravel, their form is entirely the result of rock weathering.

**55. FIGURE 7.14.** Differential weathering of Miocene marine strata at Nelson, New Zealand, has etched away erodible mudstone from between layers of siltstone, leaving siltstone standing as ridges above a wave-eroded platform.

**56. FIGURE 7.19.** A soil profile in semiarid central New Mexico includes whitish caliche forming a prominent K horizon between a yellowish-brown C horizon beneath and a reddish-brown B horizon above.

**57. FIGURE 7.22.** A thick reddish-brown paleosol (beside figure) developed on a pumice layer near Guatemala City, Guatemala, is overlain by another layer of pumice and pyroclastic flows that are separated by thinner brownish paleosols. A fault has displaced the layers about 2 m vertically.

**58. FIGURE 7.23.** Hillsides in Madagascar that have been deforested for slash-and-burn farming have been deeply gullied by erosion. Streams run red with irreplaceable topsoil that is destabilized as anchoring tree roots disappear.

59. **FIGURE 7.26.** A terraced hillside near Lanzhou, China, creates productive agricultural fields from steep hillslopes carved in deposits of windblown dust. Where unterraced, hillslopes are rapidly eroding and deeply gullied.

# CHAPTER 8
## (Slides 60–68)

60. **FIGURE 8.4.** A large slump in the high gravel terrace beside the Yakima River in central Washington has broken up a major highway and displaced it more than 100 m laterally into the river channel.

61. **FIGURE 8.6.** Talus at the base of steep cliffs in the Brooks Range, northern Alaska. When most rockfall debris moves downslope via a gully, the resulting deposit at the base of the gully is a talus cone.

62. **FIGURE 8.7.** Coarse, angular limestone blocks stand at the angle of repose (about 30°) in a talus below steep cliffs in the central Brooks Range, Alaska.

63. **FIGURE 8.10.** A meter-thick solifluction lobe has slowly moved downslope and covers glacial deposits on the floor of the Orgière Valley in the Italian Alps.

64. **FIGURE 8.11.** Debris slides that turned into debris flows on a steep vegetated mountainside in southern Puerto Rico have stripped away vegetation and inundated two houses at the base of the slope.

65. **FIGURE 8.13.** A roadcut in northern Pakistan exposes angular mudflow deposits that have accumulated as a fan along the side of a steep mountain valley.

66. **FIGURE 8.20.** A massive prehistoric debris avalanche from the northwest flank of Mount Shasta volcano in northern California left a chaotic deposit (hills in middle distance) that extends 34 km from the volcano and covers at least 450 km².

67. **FIGURE 8.22.** A jumbled mass of angular rock debris, supplied from a steep cliff in the Wrangell Mountains of southern Alaska, moves slowly downslope as a rock glacier.

68. **FIGURE 8.26.** Steep seacliffs of jointed basalt along the windward coast of Hawaii are undercut by pounding surf. When a cliff collapses, the resulting landslide debris is rapidly reworked by surf and currents, and the process begins anew.

# CHAPTER 9
## (Slides 69–76)

69. **FIGURE 9.1.** Paris, like many of the world's great cities, was founded along the banks of a major river. The Seine provides water for human and industrial use, is an avenue of transportation, and has great esthetic and recreational value. However, under the stress of a growing population, the Seine, like other urban rivers throughout the world, is increasingly susceptible to pollution.

70. **FIGURE 9.6.** During a spring flood in 1982, the White River (in upper right corner of view) inundated the small town of Jacksonport, Arkansas, which is built on the river's floodplain.

71. **FIGURE 9.9B.** Giant ripple marks formed by raging floodwaters as they swept around a bend of the Columbia River. Composed of coarse gravel, the ripples are up to several meters high and their crests are as much as 100 m apart.

72. **FIGURE 9.12.** A meandering stream near Pnom Penh, Cambodia, flows past agricultural fields that cover the river's floodplain. Light-colored point bars, composed of gravelly alluvium, lie opposite cutbanks on the outside of meander bends. Two oxbow lakes, the product of past meander cutoffs, lie adjacent to the present channel.

73. **FIGURE 9.20.** A large suspended load, eroded from extensive deposits of wind-blown silt, gives the Huang He a very muddy appearance and its English name (Yellow River).

74. **FIGURE 9.25.** Each of several terraces rising above the Waikato River in North Island, New Zealand, records a former floodplain, abandoned when the stream cut downward through underlying alluvium.

75. **FIGURE 9.26.** A symmetrical alluvial fan has been built on the margin of Death Valley, California, where a stream channel emerges from steep mountains. Light-colored sediments beyond the toe of the fan are evaporite deposits on basin flood. A braided system of channels covers the fan surface.

76. **FIGURE B9.1.** Vertical satellite view of the Aswan Dam and its reservoir, Lake Nasser. Sediment formerly carried northward to the Mediterrean Sea and now settling out in the lake will eventually fill the reservoir and make it unusable.

# CHAPTER 10
## (Slides 77–85)

77. **FIGURE 10.15.** This fissure near Mesa, Arizona, is up to several meters wide and is believed to be related to surface subsidence that resulted from the removal of large quantities of underground water.

78. **FIGURE 10.16.** The Leaning Tower of Pisa, Italy, the tilting of which accelerated as groundwater was withdrawn from aquifers to supply the growing city.

79. **FIGURE 10.21.** A marble balustrade of the Forbidden City in Beijing, China, shows the effects of more than 300 years of dissolution. The original sharply carved design has become smooth and indistinct as chemical weathering, enhanced by acid rainfall, has dissolved the stone.

80. **FIGURE 10.23.** A caver surveys the spectacular dripstone and flowstone formations of Lechuguilla Cave, a limestone cave in the Carlsbad Caverns region of New Mexico.

81. **FIGURE 10.24.** A drop of water collects at the end of a growing stalactite in Carlsbad Caverns, New Mexico. As the water loses carbon dioxide, a tiny amount of calcium carbonate precipitates from solution and is added to the end or sides of the stalactite.

82. **FIGURE 10.26.** The sacred well at Chichen Itza, a ruined Mayan city on the Yucatan Peninsula of Mexico. This cenote, formed in flat-lying limestone strata, contained a rich store of archaeological treasures that were cast into the water with human sacrifices centuries ago.

83. **FIGURE 10.27.** Cone karst near Arecibo in northwest Puerto Rico. The Arecibo radiotelescope (lower right) occupies a circular depression in the midst of a vast limestone landscape of closely spaced conical hills separated by deep sinkholes.

84. **FIGURE 10.28.** Steep limestone pinnacles surrounded by flat expanses of alluvium form a distinctive tower karst landscape around the Li River near Guilin, China.

85. **FIGURE 10.29.** The burren landscape in western Ireland is an extensive platform karst developed on limestone. Crevices have developed by solution along prominent joints, giving the terrain a regular, geometric texture.

# CHAPTER 11
## (Slides 86–98)

86. **FIGURE 11.2.** Dark bands of rock debris mark the boundaries between adjacent tributary ice streams that have merged to form Kaskawulsh Glacier, a large valley glacier in Yukon Territory, Canada.

87. **FIGURE 11.3.** A vertical satellite view of the valley-glacier complex that covers much of

Denali National Park in southcentral Alaska. Mount McKinley, the highest peak in North America, lies near the center of the glacier-covered region.

88. **FIGURE 11.4.** Several ice caps mantle the areas of highest land on Iceland in the north Atlantic Ocean. Vatnajökull, lying in the southeastern part of the island, is the largest and covers 8300 km².

89. **FIGURE 11.5.** Satellite view of Antarctica. The East Antarctic Ice Sheet overlies the continent of Antarctica, whereas the much smaller West Antarctic Ice Sheet overlies a volcanic island and surrounding seafloor. Three major ice shelves occupy large bays. The ice-covered regions of Antarctica nearly equal the combined area of Canada and the conterminous United States.

90. **FIGURE 11.17.** Surging tributary glaciers flowing from the mountains on the right of this view have deformed the medial moraines of Alaska's Yanert Glacier into a series of complex folds.

91. **FIGURE 11.18.** Recently deglaciated bedrock surface beyond Findelen Glacier in the Swiss Alps. Debris carried at the base of the glacier produced grooves, striations, and polish on bedrock as the ice flowed forward in the direction of the Matterhorn.

92. **FIGURE 11.19.** Asymmetrical glacially sculptured rock surface beyond the terminus of Franz Josef Glacier in New Zealand's Southern Alps. The glacier flowed from right to left. Slopes facing toward the glacier are smooth and polished. Scarps facing downvalley result from the plucking of bedrock blocks by flowing ice.

93. **FIGURE 11.20.** Sharp-crested arêtes flank the Western Cum, a deep cirque on the west side of Mount Everest in the central Himalaya. Sharp-crest peaks in the distance are horns that have resulted from headward growth of flanking cirque and valley glaciers.

94. **FIGURE 11.21.** Repeated glaciations by a thick valley glacier carved the deep, U-shaped Yosemite Valley in California's Sierra Nevada. The valley glacier was nourished by an extensive mountain ice cap that covered the undulating upland surface of the range, seen in the distance.

**FIGURE 11.22.** A deep fjord indents the northeast coast of Baffin Island in northeastern Canada. If the fjord were drained of water, its cross-section would resemble that of Yosemite Valley (Figure 11.21; Supplementary Slide 92).

95. **FIGURE 11.23.** A field of elongate drumlins formed during the last glaciation occupies the floor of Beagle Channel along the south coast of Tierra del Fuego in southernmost Argentina. The glacier, flowing from left to right, produced these smooth, elongate ridges of till with their long axes oriented in the direction of ice flow.

96. **FIGURE 11.24.** Till deposited by the Laurentide Ice Sheet in the James Bay Lowland of southern Canada contains numerous igneous and metamorphic clasts resting in a fine-grained matrix. The well-developed jointing probably is related to the release of stress as the 2-km-thick glacier thinned and disappeared at the end of the last glaciation.

97. **FIGURE 11.25.** Erratic boulders of Precambrian granite rest on eroded Ordivician shale on the south shore of the Gulf of St. Lawrence in eastern Canada. The boulders were carried southward across the gulf by the most recent Pleistocene ice sheet and left behind when the glacier retreated at the end of the glacial age.

98. **FIGURE 11.26.** A bouldery end moraine surrounds the terminus of Suess Glacier in Taylor Valley, Antarctica. The terminal moraine system, which rises above a frozen lake on the valley floor, passes upslope into a lateral moraine along the glacier margin.

# CHAPTER 12
## (Slides 99–112)

99. **FIGURE 12.4.** Sand ripples cross the surface of a desert sand sheet on the floor of Monument Valley, Arizona.

100. **FIGURE 12.8.** Active deflation of meltwater sediments downstream from Tasman Glacier in the Southern Alps of New Zealand produces clouds of dust. Loess is accumulating on vegetation-covered glacial deposits in the foreground.

101. **FIGURE 12.10.** A desert pavement on the floor of Searles Valley, California, consists of a layer of gravel, too coarse to be moved by the wind, that covers finer sediment and inhibits further deflation.

102. **FIGURE 12.12.** Ventifacts litter the ground surface near Lake Vida in Victoria Valley, Antarctica. The most intensely abraded surfaces are inclined to the right, in the direction from which strong winds blow off the East Antarctic Ice Sheet.

103. **FIGURE 12.13.** A sharp-crested yardang carved from compact lake sediments rises above the floor of Rogers Lake playa in southeastern California. The crest of the yardang is oriented parallel to the prevailing wind direction.

104. **FIGURE 12.16.** Sand dunes advance from right to left across irrigated fields in the Danakil Depression, Egypt.

105. **FIGURE 12.17.** A vast sand sea stretches to the horizon in the center of China's vast Takla Makan desert, a continental desert in the middle of Asia that lies thousands of kilometers from oceanic sources of precipitation.

106. **FIGURE 12.18.** A cave excavated by hand in compact loess provides a roomy and comfortable home for a Chinese family in the loess region near Xian.

107. **FIGURE 12.19.** A steep cliff of loess rises abruptly from the side of the Yellow River (Huang He) near Xian, China. The yellowish color of the river is due to the load of yellowish silt it carries, eroded from the thick and extensive loess deposits of the Loess Plateau.

108. **FIGURE 12.24.** The Pillar of God, a pillarlike butte of sandstone in the Kalahari desert of southwestern Africa, rests precariously on a pyramid of erodible shale. Erosion has separated the butte from a sandstone-capped mesa in the distance.

109. **FIGURE 12.26.** A flash flood has just passed through this steep-walled arroyo on the Navajo reservation in northeastern Arizona. As the floodwater subsides, sediment is deposited across the flat alluvial floor of the canyon.

110. **FIGURE 12.27.** A vast salt-encrusted playa occupies the floor of Death Valley in California. The playa is bordered by a bajada, individual and coalescing alluvial fans constructed beyond the mouths of adjacent mountain valleys.

111. **FIGURE 12.28.** A pediment in the Mojave Desert of southeastern California has left only a few residual hills near the crest of a former mountain ridge. The flat bedrock surface cut across crystalline rocks passes downslope beneath a thin cover of alluvium.

112. **FIGURE 12.29.** Stages in the formation of a pediment. **A.** A mountain block, uplifted along bordering faults, is eroded by streams that contribute sediment to a growing alluvial basin fill. **B.** A pediment is cut across the margins of the uplifted block and grows headward into the mountains as sheetfloods and running water transport sediment across the planar rock surface toward the basin fill. **C.** Growing headward from both sides of the upland, the pediment slowly consumes the mountains, leaving only a few residual hills rising above the eroded bedrock surface.

# CHAPTER 13
## (Slides 113–123)

113. **CHAPTER 13 OPENER.** Coastline recession, the result of storm waves pounding against a bluff of erodible sandy sediments, has undermined the foundations of this house.

114. **FIGURE 13.6.** Waves arriving obliquely onshore along a coast near Oceanside, California, change orientation as they encounter the bottom and begin to slow down. As a

result, each wave front is refracted so that it more closely parallels the bottom contours.

115. **FIGURE 13.14.** A nearly horizontal wave-cut bench has formed along the coast at Bolinas Point, California, as the surf, acting like an erosional saw, has cut into and beveled the tilted sedimentary strata.

116. **FIGURE 13.15.** An uplifted wave-cut bench at Tongue Point, New Zealand. Crustal uplift along this coast has raised the former seafloor to expose a broad bench. Light-colored beach sediment overlies the darker rocks of the wave-cut cliff at the seaward edge of the bench. Below the uplifted bench, a younger one is forming.

117. **FIGURE 13.16.** Stack and sea arch along the French shore of the English Channel near Étretat carved in horizontally bedded white chalk. The surf first hollows out a sea cave in the most erodible part of the bedrock. A cave excavated completely through a headland is then transformed into a sea arch. An isolated remnant of the cliff stands as a stack on a wave-cut bench offshore.

118. **FIGURE 13.20.** The complex spit forming Cape Cod, Massachusetts, encloses Cape Cod Bay. Waves and currents rework glacial deposits of the peninsula forming the south side of the bay and transport the sediment northward and southward. An eddy carries sediment around the north point of the spit and into the bay.

119, 120, and 121. **FIGURE 13.22.** Chief kinds of tropical coral reefs. **A.** Fringing reef on the island of Oahu in the Hawaiian Islands. **B.** Barrier reef enclosing the island of Morrea in the Society Islands. A narrow lagoon separates the high island, which is the eroded remnant of a formerly active volcano, from a shallow reef. **C.** The reef of a small atoll in the Society Islands is surmounted by low, vegetated sandy islands that lie inside a line of breakers along the reef margin.

122. **FIGURE 13.29.** Breakwaters constructed offshore from Tel Aviv, Israel protect the beach zone from incoming waves. Wave refraction around the barriers has led to progradation of the beach behind each breakwater, producing a scalloped coastline.

123. **FIGURE B13.1.** Vertical satellite image of the shrinking San Francisco Bay estuary. Filling and diking of tidal marshes to create farmland, evaporation ponds, and residential and industrial developments has reduced 2200 km² of marsh land that existed before 1850 to less than 130 km² today.

# CHAPTER 14
## (Slides 124–128)

124 AND 125. **FIGURE 14.5.** Examples of rock deformation. **A.** Fracture of strata by brittle deformation. **B.** Bending of strata by ductile deformation.

126. **FIGURE 14.7.** An orange grove in southern California planted across the San Andreas Fault. Movement on the fault displaced the originally straight rows of trees. The direction of motion is such that trees in the background moved from left to right relative to the trees in the foreground.

127. **FIGURE 14.20B.** A monocline in southern Utah that interrupts the generally flat-lying sedimentary strata of the wide Colorado Plateau. In the area of maximum bending, the strata are nearly vertical (right-hand side of photo). The view is looking south.

128. **FIGURE 14.27.** Valley and ridge province of the central Appalachians in Pennsylvania. Paleozoic-age strata have been folded into a series of plunging anticlines and synclines. Valleys are developed by erosion of the weakest strata—limestone, dolostone, and shale. The ridges are sandstones. The river in the upper-right hand corner is the Susquehanna.

# CHAPTER 15
## (Slides 129–133)

**129. FIGURE 15.6.** Refraction of light causes a drinking straw to seem to be bent at the place where it enters the water.

**130. FIGURE 15.13.** A building that was not constructed to withstand expected earthquakes. The Hotel DeCarlo was one of the buildings that collapsed during the earthquake that struck Mexico City in 1985. Proper building design can minimize damage. Nearby buildings of sturdier construction withstood the shaking.

**131. FIGURE 15.14.** When a magnitude 6.8 earthquake struck Soviet Armenia on December 7, 1988, poorly constructed buildings with inadequate foundations collapsed like houses of cards. The principal cause of collapse was ground motion.

**132. FIGURE 15.24.** Kimberlite from the Monarch Pipe in South Africa. Fragments of rock from deep in the mantle are carried upward by the forceful intrusion of kimberlite magma. Rounded fragments are the transported mantle rocks; fragmental, grayish background material is the kimberlite.

**133. FIGURE 15.31.** Beach ridges raised by postglacial uplift of the land in Nordaustlandet, Svalband, Norway. Such beaches provide clear evidence that the land has emerged from the sea after removal of the ice load.

# CHAPTER 16
## (Slide 134)

**134. FIGURE 16.1B.** Subdued landscape of Western Australia Shield with Tertiary (or possibly Jurossic) laterite at the surface.

# CHAPTER 17
## (Slides 135–141)

**135. FIGURE 17.4.** A rich vein in Potosi, Bolivia, containing chalcopyrite, sphalerite, and galena cutting andesite. The andesite has been altered metasomatically by the hydrothermal solution that deposited the ore minerals.

**136. FIGURE 17.6.** Leaching of silica during weathering of a Lake Superior-type iron deposit leads to formation of a secondarily enriched mass of iron minerals that is rich enough to be an ore. This sample, from the Hamersley Range in Western Australia, was developed by secondary enrichment from the kind of iron-rich sediment shown in Figure 4.14 (Supplementary Slide 33).

**137. FIGURE 17.7.** Stratabound ore of lead and zinc from Kimberley, British Columbia. The layers of pyrite (yellow), sphalerite (brown), and galena (grey) are parallel to the layering of the sedimentary rock in which they occur. The specimen is 4 cm across.

**138. FIGURE 17.9.** Gold is recovered from ancient fossil deposits of the Witwatersrand, South Africa. The gold is found at the base of conglomerate layers interbedded with finer-grained sandstone, here seen in weathered outcrop at the site where gold was first discovered in 1886.

**139 and 140. FIGURE 17.10.** Residual mineral deposits rich in iron and aluminum are typically formed under tropical or semitropical conditions. **A.** Red laterite enriched in iron, near Djenné, Mali. The upper two meters of the laterite consist of rounded concretions of limonite cemented by more limonite to create a hardened mass. Residual iron

ores have been mined in the past, but no large mining activity is occurring today.
**B.** Bauxite from Weipa in Queensland, Australia. Long-continued leaching of clastic sedimentary rocks under tropical conditions has removed all original constituents, such as silica, calcium, and magnesium, leaving a rich bauxite consisting largely of the mineral gibbsite ($Al(OH)_3$). Nodules of gibbsite form by repeated solution and redeposition. The Weipa bauxite deposits are among the largest and richest in the world.

141. **FIGURE 17.17.** A field of windmills near Palm Springs, California, designed to generate electricity using the kinetic energy of wind.

# CHAPTER 18
## (Slides 142–149)

142. **FIGURE 18.2.** Multiple concentric rings around Valhalla, an impact crater on Callisto, a moon of Jupiter. The crater (central bright spot) is 600 km in diameter. The rings were formed as a result of thrust faults caused by the compression of the giant impact. The surface of Callisto is ice. Rings are also observed around giant craters on rocky planetary bodies, but they are neither so pronounced nor so numerous as the rings around Valhalla.

143. **FIGURE 18.3.** A large impact crater on the Moon. The crater is more than 200 km in diameter; the oblique photo was taken from a manned spacecraft. The highlands in the far distance are part of the ejecta blanket. The stepped terraces in the middle distance were formed as a result of postcratering collapse of the crater rim. The hills in the foreground lie at the center of the crater and were formed by rebound of the crater floor during decompression.

144. **FIGURE 18.7.** A photograph of the surface of a mare. Taken by an astronaut during the *Apollo 11* mission, the photograph shows portions of Mare Tranquillitatis. The irregular ridges, looking like long sand dunes, are ancient basaltic lava flows. The impact crater on the lower right-hand side is called Maskelyne.

145. **FIGURE 18.10.** Olympus Mons, a giant shield volcano on Mars, is the largest volcanic edifice discovered so far in the solar system. The image is a computer-enhanced view made from a number of images taken by *Mariner* spacecraft.

146. **FIGURE 18.11.** Detail of the summit caldera of Olympus Mons. Several phases of caldera collapse have occurred. The irregular ridges in the center foreground are basaltic flows.

147. **FIGURE 18.13.** An image of the Martian surface from *Viking 2.* The red color indicates that the surface rocks have undergone oxidation as a result of weathering.

148. **FIGURE 18.15.** Steep-sided lava domes on Venus seen by the *Magellan* spacecraft early in 1991. Each dome is about 25 km in diameter. The shapes and steep sides resemble rhyolitic lava domes on the Earth, and this suggests that high viscosity magma might be produced on Venus by some form of magmatic differentiation.

149. **FIGURE 18.18.** The surface of Ganymede, largest of Jupiter's moons, viewed from a distance of 312,000 km by *Voyager 2* in 1979. Ganymede is covered by a thick crust of ice. The dark surface is ancient ice, presumably covered by dust and impact debris. It is split into continent-sized fragments that are separated by light-colored, grooved terrains of younger ice. Ganymede is apparently tectonically active, and the grooved terrains seem to be the places where new ice rises from below, but how this happens, and what causes the grooves, is not known. The field of view is approximately 1300 km across.

# CHAPTER 19
## (Slides 150–151)

**150. CHAPTER 19 OPENER.** This NASA false-color image shows average total ozone values in the South Pole region during October 1987. The dark purple color indicates extremely low ozone values, whereas the yellowish colors indicate high values.

**151. FIGURE 19.5.** Air bubbles trapped in glacier ice. By melting the ice in a laboratory and collecting the gas, the content of $CO_2$ and other trace gases in these ancient samples of the atmosphere can be measured.